Television in the Multichannel Age

A Brief History of Cable Television

Megan Mullen

Blackwell
Publishing

BLACKWELL PUBLISHING
350 Main Street, Malden, MA 02148-5020, USA
9600 Garsington Road, Oxford OX4 2DQ, UK
550 Swanston Street, Carlton, Victoria 3053, Australia

The right of Megan Mullen to be identified as the author of this work has been asserted in accordance with the UK Copyright, Designs, and Patents Act 1988.

Designations used by companies to distinguish their products are often claimed as trademarks. All brand names and product names used in this book are trade names, service marks, trademarks, or registered trademarks of their respective owners. The publisher is not associated with any product or vendor mentioned in this book.

This publication is designed to provide accurate and authoritative information in regard to the subject matter covered. It is sold on the understanding that the publisher is not engaged in rendering professional services. If professional advice or other expert assistance is required, the services of a competent professional should be sought.

First published 2008 by Blackwell Publishing Ltd

1   2008

Library of Congress Cataloging-in-Publication Data

Mullen, Megan Gwynne, 1964-
   Television in the multichannel age : a brief history of cable television /
Megan Mullen.
      p. cm.
   Includes bibliographical references and index.
   ISBN 978-1-4051-4969-3 (hbk. : alk. paper) — ISBN 978-1-4051-4970-9
(pbk. : alk. paper)   1. Cable television—History.   I. Title.
   HE8700.7.M85 2008
   384.55'5—dc22

                                                      2007024788

A catalogue record for this title is available from the British Library.

Set in 10.5/13pt Galliard
by Graphicraft Limited, Hong Kong
Printed and bound in Singapore
by C.O.S. Printers Pte Ltd

The publisher's policy is to use permanent paper from mills that operate a sustainable forestry policy, and which has been manufactured from pulp processed using acid-free and elementary chlorine-free practices. Furthermore, the publisher ensures that the text paper and cover board used have met acceptable environmental accreditation standards.

For further information on
Blackwell Publishing, visit our website at
www.blackwellpublishing.com

Television in the Multichannel Age

In memory of my mother, Sally Jones Mullen

Contents

Illustrations

Preface

In 2007, cable and satellite television are at a crossroads. In order to survive much farther into the new millennium, these industries and the technologies they rely on must find ways to adapt to, even merge with coexisting media such as telephony, networked computers, and other digital technologies. This is already in process. Cable companies now routinely offer their customers high-speed Internet service – even while that service provides access to streaming video and other Internet-based media that compete with cable's own programming. Major telecommunications providers, already offering high-speed Internet, are poised to offer cable-type television programming via that same medium very soon. And, hedging their bets in an uncertain future, telecommunications companies have begun purchasing or merging with direct broadcast satellite providers. These efforts are taking place worldwide, since today's media markets routinely extend across national, even continental boundaries. Policy-makers virtually everywhere are contending with this reality, as media structures designed for individual nations are constantly challenged by incursions from abroad.

Yet it seems to me that, even though cable and satellite television are on the verge of losing their autonomy in the entertainment and information universe, very little has been recorded about the specific histories that have brought them to this point. So this is a book that had to be written. Although I had some other projects in process at the time of being granted a sabbatical leave for 2005–6, I knew this book would consume the bulk of my time away from teaching and service obligations. While there are many things I still need to capture in print about specific periods and phenomena in cable history, I felt a sense of urgency in producing a book that would be read by people other than my fellow academics. I wanted to write something for undergraduate students, employees of cable and related industries, and laypeople generally.

I've been fortunate with my resources. Although comprehensive volumes dealing with multichannel television history are few, the body of academic scholarship on cable programming, policy, technology, and history is growing every day. There also have been extensive efforts on the part of the cable industry and others to capture and transcribe oral histories and archive files and equipment. I am especially indebted to the Cable Center, located in Denver since 1998 and at Penn State University prior to that, for the availability of these resources and capable individuals to guide me through them. In particular, I wish to recognize the assistance of Lisa Backman (Denver), Beth Lazarczyk (Denver), and Pamela Czapla (Penn State).

Beyond archived materials, individuals who remember cable's past are an extraordinary resource. In my head I now hold many stories told to me by individuals who have worked in the cable television industry at various points during its six-decade history. Many of these individuals will not live to see what lies in cable's future five or ten years from now; in fact a few passed away while I was writing this book. I owe it to them to capture the history they lived through – a history they shared with me in vibrant detail – to the best of my ability. So I dedicate this book to those who have worked in the cable industry who have provided so much assistance, not just with this project but throughout all my efforts to learn and record cable history. I especially recognize my friend E. Stratford "Strat" Smith, who has spent many hours over the past couple of decades sharing his own recollections with me, helping me to understand several of the legal issues in cable television history, and connecting me with other key individuals in the industry.

The recollections of some of those individuals had a direct impact on this book, in that they drew my attention to places and events previously unknown to me. These individuals include James Duratz, James Strickler, and the late Yolanda Barco of Meadville, Pennsylvania; the late Robert Tarlton of Lansford, Pennsylvania; James Y. "Jimmy" Davidson of Tuckerman, Arkansas; and Martin F. Brophy of Shenandoah, Pennsylvania. The book was also informed by the recollections of individuals I interviewed for my first book and who are acknowledged therein.

For this particular project, I received generous financial support in the form of a year's sabbatical leave from the University of Wisconsin-Parkside and a National Endowment for the Humanities Faculty Fellowship.

I also received practical assistance from individuals including Harland Bergstrom, Cable Services Company (assistance with technical illustrations); Michelle Curtis, RLJ Companies (Robert Johnson photo and editorial suggestions); Jenel Farrell, National Public Radio (permission to use the transcript of the Ted Koppel interview); Lana Khachan, Al Jazeera English (Doha studio photo and editorial suggestions); Debbie Lamb, C-SPAN (Brian Lamb photo); Betsy Lambert, CommScope (photos of coaxial and fiber-optic cable); Ruth Lee, ATX (photo of headend interior); Megan Milner, Moonview Sanctuary (photo of Gerald Levin); Rachelle Savoia, Oxygen (Geraldine Laybourne photo); Pam Steitz and Rachel Packard, Sun Prairie Cable Access (photo of KIDS-4 production); Elisa Travisono, UN Foundation (permission to use Timothy Wirth photo and editorial suggestions); and Angell Vasko, CBN (permission to use Pat Robertson photo).

Peer reviewers for this manuscript provided extensive and helpful suggestions, for which I am most grateful. My friend and former professor Sharon Strover provided advice along the way and deserves recognition for introducing me to complex policy terrain around the cable and satellite industries when I was her student. I acknowledge colleagues and administrators at the University of Wisconsin-Parkside who gave help and encouragement along the way. Finally, I wish to thank Ken Provencher and Jayne Fargnoli at Blackwell for both their extraordinary capabilities and their ongoing words of encouragement. They made writing this book an enjoyable experience from the start.

Chapter 1

Introduction

Many Americans born after 1980 have trouble distinguishing between broadcast and cable/satellite television channels, especially if they have lived in cable- or satellite-subscribing households all their lives. They can't remember a time when "channel surfing" meant switching among three, four, or five channels – all of which had more or less the same types of programs at the same times of the day and week. We take multichannel television for granted today; even if we don't subscribe to cable or direct broadcast satellite (DBS) services ourselves, we've surely heard of the content available through these media technologies. And although they sometimes don't like to admit it, most non-subscribers, at one time or another, have wished to see a non-broadcast program. This is hardly surprising. The array of television programs available through cable, DBS, and even the Internet in the twenty-first century is at least a hundred times greater than what is available using a broadcast antenna alone. There are channels to suit most people's interests. And where people once complained that "nothing" was on, now they are more likely to complain of not being able to choose from a mind-boggling array of possibilities.

Cable television has not always been a multichannel supplement to broadcast television, however. Cable actually was "invented" in the United States because of a combination of geographical conditions and regulatory uncertainty. Many initially perceived it as a temporary, makeshift technology. Yet such disparate factors as an extraordinarily cohesive trade organization in the early years, a fickle regulatory climate that called for a great deal of individual ingenuity, and the parallel development of complementary technologies such as communications satellites and enterprises such as pay-television all helped to mold cable into a distinct and prominent media industry. Thus the project of this book is to weave together the varied threads that offer an account of how multichannel television came to be what it is today – both in the United States and throughout the world.

Until the late 1970s, cable's almost exclusive function was to re-transmit broadcast television signals to communities where residents would have been unable to receive those signals using home antennas (whether the set-top "rabbit-ear" antennas or rooftop antennas). All broadcast signals can be received relatively clearly 60–100 miles from a broadcast station's transmitter – provided there are no obstacles such as mountains or skyscrapers. However, many communities *do* need to contend with these sorts of obstacles blocking television reception. Moreover, many communities in the United States are simply too far away from broadcast stations to receive their signals at all. Thus cable television got its start in the United States as a means of bringing television to these sorts of communities by using a very tall antenna, known as a community antenna. Community antenna television, or CATV, was the first name for cable television.

Additionally, CATV began during the FCC's 1948–52 licensing "freeze," years in which policy-makers suspended the assigning of new television frequencies in order to examine their practices for allocating licenses to prospective new television stations. This was ostensibly to ensure equal availability of television service nationwide, regardless of community size or remoteness from centers of popula-tion. During this time television was known to virtually all Americans, even though its availability was very uneven: a few areas of the country were very well served by television while many others lacked recep-tion entirely. In a number of communities, particularly on the East coast, CATV systems were built to extend the reach of existing tele-vision stations. The technology was considered benign by policy-makers, most seeing it as nothing more than a stopgap technology that would disappear once more television stations had been granted licenses and begun service. Of course this never happened, as we now realize.

Since no widespread alternatives to commercial support emerged during US television's early years, only medium and large cities have the populations to sustain television stations of their own. The four-year licensing freeze was just long enough for the fledgling CATV industry to gain a foothold. There continued to be a role for that industry long after the freeze had ended; in fact CATV's role even grew due to the post-freeze granting of frequencies in the UHF (ultra high frequency) band – which were much weaker than those in the VHF (very high frequency) band. By the mid-1960s, CATV was desired even in communities with one or two stations of their own, since community antennas (and related technologies) made it possible to bring in additional channel choices.

In the ensuing decades, new technologies – most notably communications satellites – and changes in policy brought about even more enhancements to CATV's basic function. The medium became known as *cable* television and began to offer channels that were not available from broadcast television stations at all. From the 1970s through the 1990s, many dozens of satellite-delivered cable networks came on the scene and changed what was essentially a broadcast-enhancement technology into a form of television all its own.

By the early twenty-first century, it is no longer possible to consider the broadcast and cable/satellite television industries in isolation from each other. Even those households that still receive television exclusively via home broadcast antennas are at pains to avoid series or movies originally produced for cable/satellite networks. Popular cable/satellite networks such as CNN can be viewed in airports and other public places. And icons from popular cable programs (e.g., SpongeBob SquarePants or the Soprano family) circulate freely within popular culture. Moreover, cable technology, now merged with computer and telephone technologies, offers consumers an array of services beyond simply television programming – including high-speed Internet access, digital telephone services, and video-on-demand (VOD).

Today's multichannel television is hardly limited to the context of the United States. In fact, at every stage of US cable/satellite history there have been related – sometimes even intertwined – developments in other nations. Two different trajectories need to be pursued in this regard: the one in which nations other than the United States have discovered and developed television-related technologies suited to their unique needs and the one in which the popular, successful, and controversial US-derived model of television service has made its way into other national media scenarios.

The challenge in capturing and relating the history of today's multichannel television, therefore, is to determine how many different "histories" to draw into the picture. Not only must we consider the existing media landscape into which a new technology enters; we must also look at the ways in which the capabilities of the new technology play a role in the future of that scenario. All media technologies are affected by engineering breakthroughs, government policy at all levels, economic fluctuations, and dominant ways of doing business. There are also considerations related to the individuals involved in the industry – ranging from risk-taking entrepreneurs to innovative program producers. And the audience cannot be overlooked, for even the most groundbreaking business or programming innovations can succumb to the whims of fickle and often conservative consumers.

What I offer here is a historical overview of these industries. I do not intend to treat any aspect of cable/satellite history exhaustively or in a way that would add a great deal to a media historian's existing expertise. Rather the book is aimed at the beginning-to-intermediate media student, someone working in the media industry, or the general reader. I've attempted to compile this history in as comprehensive, representative, and interesting a way as possible. I've drawn from a range of primary and secondary sources: published histories of broadcast media, books specifically on cable (including some very good histories for the time periods in which they were written), surveys of national media industries, government documents, trade press articles, policy studies, biographies, oral history transcripts, and personal interviews.

While a number of books have been published on the cable industry over the years, these have tended to be biographies or autobiographies, journalistic accounts, or visionary treatises. Though useful in most cases, especially for capturing individuals' motivations and personality factors, most of these books are both out of date and out of print. There are some more recent academic books on cable. For example, Parsons and Frieden's *The Cable and Satellite Television Industries* (1998) offers a detailed overview of cable/satellite technology, policy, and industry structures. However, only one chapter covers history specifically. Lockman and Sarvey's *Pioneers of Cable Television* (2005) provides valuable discussions of selected industry pioneers from Pennsylvania. Laura Linder's *Public Access Television* (1999) looks at one important and often overlooked part of cable television history. And there are a number of scholarly articles as well as book chapters that contribute to the understanding of multichannel television history. The history of multichannel television clearly is a growing area of scholarly interest. And yet there is still a role for a book that surveys the range of topics to be pursued.

When I first started researching cable television history a little less than two decades ago, the challenges were different than what they are today. It was definitely easier to keep track of developments in the industry, programming innovations, and new technologies. The industry seemed very large back then, but it was tiny by today's standards. New programs and programming strategies were easy to observe, record, and analyze on the few dozen cable networks that came with my subscription (and perhaps a dozen more that were available where my parents lived). And cable television had barely begun to intertwine with other media and telecommunications industries.

But it was much harder at that time to research the industry's history. I used inter-library loan services extensively and made numerous trips to archives – or even to visit some of the actual cable pioneers who were still living at that time (most of them no longer are). Today, most of the out-of-print books are available via the Internet from secondhand booksellers. And a portion of the archival material I use is available online. Convenience notwithstanding, I must admit that I miss all of the travel, not to mention the excitement of meeting the individuals who "made" the history I study. Sometimes I still visit what I call the "early cable towns" – places such as Shenandoah, Lansford, Pottsville, and Schuylkill Haven in eastern Pennsylvania's anthracite coal region; Meadville in western Pennsylvania; and of course my hometown, Oneonta, New York. I should point out that all of these communities (and many others) are tied up with my own personal history, many being places where either I or my ancestors have resided, and this has been a major draw for me in researching this area of media history.

Outline of the Book

The purpose of this book is to examine the history of multichannel television, as we know it in the twenty-first century, beginning with its origins in the late 1940s and continuing through today's range of television-related technologies. For the most part, the book focuses on the United States specifically, although each chapter also includes a brief survey of developments in cable and other multichannel television technologies in other nations throughout the world at that stage of history. The latter is intended not to provide a comprehensive overview of multichannel television around the world – a project yet to be taken on – but rather to offer a sense of how the multichannel television industry in the United States compares with, affects, and intersects with those in other nations.

The remainder of Chapter 1 looks at how cable, satellite, and related technologies function. It considers the structures of multichannel television industries and how those industries make money. And it examines the various sources of programming available through one or more forms of multichannel television.

Chapter 2 begins with the pre-history of US cable television, including the rise of the telegraph as the first electronic communication medium in the nineteenth century, radio's transition from "wireless telegraph" to a broadcast medium, and the beginning of broadcast television. This material is included to help explain the technology and policy precedents inherited by the cable industry as well as the need for its advent in the first place. The chapter then looks at the origins of cable in the late 1940s and 1950s. This was the community antenna phase of multichannel television's history. During these years, cable television was highly experimental, as small-town entrepreneurs tinkered with a variety of technologies in order to bring television to communities where its broadcast signals could not be received over the air. It was also during this time that a distinct CATV industry coalesced and formed the trade association that still dominates the industry today. Very little government regulation of CATV was undertaken during the 1950s (even though most realized it was inevitable in the future), leaving the emerging industry with a fair amount of latitude in defining itself.

The early to mid-1960s proved quite a contrast – as will be discussed in Chapter 3. Federal government policy-makers became much more involved with CATV at this point, their stated main concern being how to keep CATV systems from usurping the fortunes of fledgling broadcast television stations (especially those in smaller markets). It was much less costly to operate a community antenna system than a broadcast television station. And CATV was proving to be at least as well received by the public, especially as new technologies such as microwave relays increased the variety of channels available to subscribers. The 1960s saw the implementation of policy that severely restricted the conditions under which CATV could operate in the United States. Meanwhile, a handful of CATV-type services were starting to flourish elsewhere in the world, particularly in Canada.

Chapter 4 covers the years from 1968 to 1974, when CATV started to be known as cable television. Oddly enough, during these years US policy-makers made an almost 180-degree reversal of the stance taken earlier in the decade. No longer fearing that cable would put broadcast television out of business, they (along with various other constituencies) began to cast cable as the potential rescuer of local, educational, and special-interest television programming. This was the era in US cable history known as "Blue Sky." These years saw the rise of local public-access channels on cable, as well as pay-cable services – most notably Home Box Office – that would develop into today's satellite-carried cable and DBS networks.

Chapter 5 covers the rise of commercial communications satellites and the non-broadcast-derived cable services (networks) they helped bring about. At this point, US cable was starting its transformation from a retransmission technology into a multichannel supplement to broadcast television. These were deregulatory years for cable, and while the concept of local public access did not disappear entirely, it became clear that what most of the US public wanted to watch (and pay extra money for) were cable channels that brought more of the sorts of programs and genres that already had a home on broadcast television's "big three" networks (ABC, CBS, and NBC). For a variety of reasons, by this point there was more television programming flowing across national borders. Of course the United States was sitting comfortably in its role as the overwhelmingly dominant exporter of television programs, but other nations' programs (particularly from the British Commonwealth) were becoming more available to American audiences.

By the 1980s and early 1990s, as covered in Chapter 6, there were many more satellite-delivered cable networks, including such "standards" as CNN, MTV, and ESPN. For this reason, markets already well served by broadcast television were being wired for cable. Large amounts of money were at stake as cable companies vied for exclusive rights to provide service to these cities. The cable industry, once a small-town operation, was by this stage largely under the control of the same entertainment industry corporations that dominated other media industries. Cable was starting to play a major role in the increasing synergy enjoyed by those corporations. The federal government, especially Congress, was also taking notice. The 1984 Cable Act was the first piece of legislation to deal specifically with cable. It would be followed quickly by the very different 1992 Cable Act. Meanwhile, competition for the cable companies was emerging in the form of, first, home receivers for the signals of satellite networks and, later, DBS. The former were the very large dishes people would put in their yards, while the latter are the small dishes people now attach to the sides of houses or apartment buildings. By the end of the 1980s both cable and DBS companies were also gaining a foothold in countries other than the United States.

Chapter 7 deals with worldwide developments in cable and DBS during the 1990s and the early part of the twenty-first century. By this stage, particularly in the Unites States, most satellite networks begun in the 1970s and 1980s had recovered start-up costs and were able to devote more resources to original programming. Multichannel television was again on the agenda of the US Congress, with the far more comprehensive (and deregulatory) 1996 Telecommunications Act. By the end of the decade, cable and DBS were in head-to-head competition, with other technologies, both complementary and competing, on the horizon. Also by the end of the decade, nations throughout the world were contending with the persistent, large-scale influx of US-style multichannel television businesses.

By the early twenty-first century, it would be clear that the key to the cable industry's future success lay in its ability to integrate its technological and business structures with those of other media, telecommunications, and computer industries. As will also be discussed in Chapter 7, a variety of telecommunications technologies, including satellites, contributed to the collapse of the Soviet bloc during the late 1980s and early 1990s. Satellite transmissions have also presented huge challenges to communist governments in nations such as China and Cuba.

Chapter 8 then provides a reflection on the material covered in this book – from cable television's community antenna years through the era of satellites, Internet, and other converging multichannel television technologies. This final chapter also invites the reader to engage in some critical speculation about the future of multichannel television in the United States and throughout the world.

How Cable Television Works

The first step in understanding how *cable* television works actually involves understanding how *broadcast* television works, since retransmitting broadcast television signals has been cable's first and most prominent function. Broadcast television – like its predecessor, broadcast radio – relies on a powerful transmitting antenna to send programs through the airwaves to area households, which then access that programming using a household receiving antenna (either a set-top "rabbit-ear" antenna or a larger rooftop antenna). The receiving range for a transmitting antenna is typically 60–100 miles, depending on both signal strength and terrain. The programming transmitted is the product of local stations, which have either acquired it specifically for this sort of distribution or produced it themselves.

The next step in understanding cable television is the technology that enables delivery of the broadcast signals discussed above beyond the reach of the transmitting antenna – as well as a host of complementary technologies that allow cable to offer more than broadcast signals alone. Since its very beginning, cable's central component has been the headend. This is the site (often on top of a community's highest hill or mountain) where all signals are gathered. In the early days, the headend was nothing more than a large tower with antennas on top (one for each broadcast signal to be received). By the early 1960s, many headends also included microwave receiving dishes (typically attached to the antenna tower) to enable reception of more distant signals via microwave relay. Microwave is a point-to-point transmitting technology. In other words, a broadcast signal is "bounced" from one transmitting tower to another until it reaches the place(s) where it will be used. Because microwave is a "line of sight" technology, meaning that signal strength diminishes over distance due to the earth's curvature, multiple transmitters are needed along the route – which can range from around 50 miles to over 1,000 miles in length.

In the 1970s, satellite dishes (also known as earth stations) for receiving satellite-carried program services or networks became part of many headends as well. The satellites that carry cable network signals are in geosynchronous orbit 22,240 miles (35,784 km) above the earth's surface. At this precise distance, the satellites remain positioned above the same area of the planet, providing continuous signals to that area. This area is called the satellite "footprint" and its size varies depending on the receiving and transmitting technologies. Any dish located within the footprint can receive the satellite's signals. Signals of individual networks are carried on portions of satellites called transponders.

A headend also has a building where signals of all origins are processed (i.e., cleared of interference, amplified, and sometimes descrambled). In some communities this building is located near the antenna tower, as it was when cable first began, but increasingly it is located adjacent to the system's business office, with a direct line feeding signals from the various antennas. This building also houses various playback technologies (videocassette players, DVD players, etc.) so that any locally produced programming can be

Figure 1.1 Headend – exterior. Aside from the presence of satellite receiving dishes, many of today's mountaintop headends look much the same as they did in the 1950s [author photo]

Figure 1.2   Headend – interior. Although headend buildings are seldom elaborate on the outside, inside they contain an array of sophisticated processing and transmitting equipment [ATX]

added to the line-up of channels going out to subscribers. In recent years it has also begun to house technologies related to other services offered by cable companies, such as broadband Internet service, cable DVRs (digital video recorders), and cable telephone service.

After signals have been processed at the headend, they are sent into the community via the trunk, a line of either coaxial or fiber-optic cable. The trunk then carries the signals to feeder lines, which extend into quadrants or neighborhoods in the community. The line of cable that extends from the feeder to a cable subscriber's home is called the drop. On the way from the headend to subscribers' homes, the cable signals need to be amplified multiple times owing to attenuation or signal fading. Over the years, cable engineers have developed increasingly sophisticated amplifiers for this purpose. Another technology along the modern-day cable line is the tap, a device that controls which signals a subscriber will receive. It may or may not contain a descrambler. Traditionally the lines of cable were carried on utility poles. Nowadays they, like other utility lines, are often run through underground tubes called conduits.

Inside a subscriber's home, a converter box is often needed. Most of cable's analog (i.e., non-digital) channels are converted at the headend into frequencies outside the VHF and UHF bands used by broadcast television. This allows more channels to be carried, and since cable is enclosed, there is no need for concern over interference with other over-the-air technologies (e.g., police radio) that use those frequencies. Historically, converter boxes either have allowed cable signals to be tuned using the VHF dial or – when more than twelve cable channels are available – have provided an alternative tuning device. Today, virtually all television sets sold are "cable-ready," meaning that they are capable of tuning all the frequencies used by cable. Many subscribers no longer need converter boxes. However, increasingly converter boxes are being used for new services such as digital channels, interactive services, and digital video recording. Cable-related services

Figure 1.3 Most types of transmission cable appear similar on the outside [Getty Images]

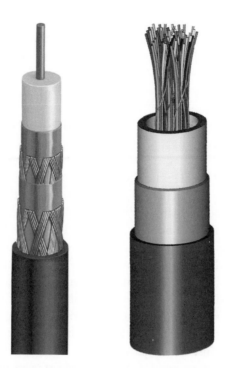

Figure 1.4 Inside configurations vary quite a lot. On the left is coaxial cable; on the right, fiber-optic cable [CommScope, Inc.]

such as high-speed Internet and telephone service add even more varieties of "cable boxes" to subscribers' homes.

The cable wire itself has also changed over the years. In the earliest days of community antenna television, twin-lead or ladder-lead cable was used. This consisted of two copper wires, usually connected by plastic. Twin-lead cable had several problems in outdoor use, however, including a tendency to lose signal during rain. In the community antenna industry, twin-lead was replaced (before the vast majority of systems even started) by coaxial cable. Coaxial cable (known to insiders as "co-ax") is much more weather-resistant. It consists of a copper wire at the core, surrounded by foam, which itself is surrounded by woven copper. The woven copper is again surrounded by foam, and the entire cable is covered with plastic. While coaxial cable is still used extensively, it is increasingly being replaced by fiber-optic cable, which consists of numerous flexible glass threads wrapped in a plastic casing. Because fiber-optic cable uses pulses of light for transmission, thereby maximizing the amount of information that can be transmitted, it has made a number of new services available to cable subscribers. It is particularly well suited to digital programming and data services.

How Direct-to-Home and Direct Broadcast Satellite Work

Direct-to-home (DTH) satellite is a broad term referring to any satellite-delivered multichannel program service that reaches subscribers via a home receiving dish instead of cable service. The concept of DTH originally took hold in the early 1980s. At that point, some individuals – especially those living in isolated areas not passed by cable – realized that they could receive the same program networks as cable operators by erecting home receiving dishes. These were the 10–12-foot C-band earth stations that cable systems themselves were using to receive signals of the satellite-carried program networks they were delivering to subscribers along with broadcast signals. At first these home satellite mavericks seemed to be getting something for nothing, but as more and more people started doing this, cable networks began scrambling their signals and businesses sprang up to sell cable-type subscriptions (with descrambling technology) to home satellite users. As of 2007, C-band DTH has only a handful of remaining subscribers in rural areas of the United States, though the technology is used more extensively in other countries.

More common today is direct broadcast satellite (DBS): the 18–20-inch dishes that can be attached to the side of virtually any type of house, apartment, or other building. This technology transmits on the Ku band of the electromagnetic spectrum instead of the C band. The smaller wavelengths of the Ku band allow a signal to be focused on a much smaller point. When the licensing of Ku-band DBS was announced by the FCC in the early 1980s, there were thirteen applicants, eight of which were granted licenses. This was the number of orbital slots that had been allocated for this use. However, most of the original licensees, along with other aspiring providers who subsequently applied, either could not meet the initial start-up costs or went out of business shortly after launching. Today, two competing brand-name technologies and their accompanying program subscription services, DirecTV and Echostar (DISH Network), vie for virtually all the DBS subscribers in the United States.[1] A challenge faced by these companies, as well as others that tried to launch in the 1990s, had been their inability to provide the signals of local broadcast stations – still the most popular channels on television. This situation was rectified by an FCC requirement, effective January 1, 2002. Both DirecTV and DISH now offer subscription packages that are competitive with cable service and are taking away a growing share of cable's market.

Other Multichannel Television Technologies

It should be stated up front that cable and DBS overwhelmingly dominate the multichannel television marketplace both in the United States and internationally. However, there are a few other multichannel options that should be mentioned. First is SMATV (satellite master antenna television), a system in which multichannel programming is transmitted by satellite and received by a dish that serves an entire hotel, motel, hospital, or apartment building. The signals are transmitted from the receiving dish via coaxial cable. Second is MMDS (multichannel multipoint distribution service), also known as "wireless cable." MMDS is essentially a way of distributing cable programming without actual cable connections. MMDS uses microwave technology for transmitting programming from satellite receiving dishes via a headend-type facility to subscribers' homes. In recent years, potential users of MMDS – primarily households in areas not passed by cable – have opted instead for the more efficient DBS delivery systems. Other multichannel delivery systems are certain to emerge in the near future. For example, as Internet technologies increasingly merge with cable and satellite services, we are beginning to see new options for multichannel video delivery, including Internet-based streaming video services and the digital multichannel services being introduced by telephone companies.

Types and Sources of Multichannel Programming

Virtually every kind of multichannel television provider offers a variety of channel types. As discussed above, the most prominent type of channel historically has been the retransmitted broadcast station. But in the mid-1970s, with the rise of satellite technology, cable-specific program services or networks began to form as well. These include basic cable networks such as CNN, MTV, and ESPN that are "bundled"; in other words, subscribers are charged a flat monthly fee to receive several of them. With basic cable service, most networks are compensated by cable or satellite operators on a per-subscriber or "per-sub" basis, meaning that a fee from a few cents to a few dollars per channel is folded into each subscriber's monthly bill. With bundling, subscribers wind up paying for channels they seldom or never watch. Thus there has been a great deal of debate as to whether basic networks should be offered on an individual or "à-la-carte" basis. This change has almost never been implemented, however, because doing so would be very likely to put less popular or special-interest networks out of business.

Another type of cable/satellite channel is the pay-cable or premium channel. This category includes networks such as HBO and Showtime, for which subscribers pay additional monthly fees that average $20–$25 per channel. Nowadays, a subscription to a premium channel typically includes access to more than one satellite feed (e.g., East Coast and West Coast) or sub-service (e.g., HBO Signature, HBO Comedy). There also are pay-per-view (PPV) channels that allow subscribers to pay directly for individual movies or special events (such as sports events). Traditional PPV, which required watching the program at a specific time, increasingly is giving way to "on-demand" PPV, which allows a viewer to choose when to watch a program (typically within a 24-hour period), as well as to start and stop the program as if it were being viewed on videocassette or DVD. On-demand PPV is made possible by the digital storage and transmission of video recordings. Another increasingly common part of digital cable packages are digital music channels.

Today, both cable and DBS providers offer multiple levels or "tiers" of programming selections. At cable's most basic level, it is possible to receive only broadcast stations along with a few popular cable networks and any public access channels available in the community. People who watch little television but receive no clear signals over the air might opt for this level of service – even if only to receive local newscasts. This tier is inexpensive and not heavily promoted. More people subscribe to the next level of service, which for most cable systems consists of broadcast signals and public access, as well as around 50–100 basic cable networks (such as CNN, MTV, and USA). In the twenty-first century, most cable systems now offer digital service packages of various configurations as well. These include everything listed above, along with considerably more basic cable channels, all-music channels, and video-on-demand (VOD) channels. Digital services take advantage of increased bandwidth (i.e., the amount of signal that can be carried) made available by new technologies such as computer-operated cable boxes and fiber-optic cable. This is the level at which cable is most competitive with the multichannel array offered by the two major DBS providers.

Broadcast signals continue to be a popular source of programming in the multichannel age. While the look of broadcast-derived channels on cable and satellite is identical to that of other types of channels, it is important to note that they come from a very different source. Broadcast signals originate at broadcast stations located in medium to large cities. Their coverage is regional, extending as far as the signal can travel without fading or interference. A broadcast station in the United States is identified by a set of three or four call letters, beginning with "W" (mostly east of the Mississippi River) or "K" (mostly west of the Mississippi), that represent a license granted by the Federal Communications Commission (FCC). Today, many broadcast stations are better known by "nicknames" such as "News 12-Springfield" or "ABC 7-Chicago."

A typical broadcast station is either affiliated with a national broadcast network (most commonly ABC, CBS, NBC, or FOX) or independent. If it has a network affiliation, it has agreed to air the programs distributed by a particular network in exchange for financial or other compensation. This is beneficial to the station since national network programming generally is the most popular, heavily promoted, and lavishly produced. And since the networks leave some commercial time for the local affiliate, that station can use the popular programs to draw local advertising revenues. The stations do not show network programming throughout the entire day, however. They also show local news and other locally produced programs.[2] If the station has no network affiliation (as is seldom the case anymore), it must rely on other sources of programming to fill its day. In a few cases, independent stations in large cities enjoy a great deal of success as outlets for major sports events (e.g., New York's WOR, Chicago's WGN, and

Los Angeles's KTLA all serve as television "homes" for their local Major League Baseball teams).

Perhaps even more significantly, however, both network affiliates and independent stations, as well as many basic cable networks, rely heavily on syndicated programs – in the form of old movies, television reruns, and newly produced programs (such as some daytime talk shows or game shows). Syndication refers to the practice of distributing programs on a market-by-market basis instead of through an affiliation agreement. The station or cable network pays a syndication company or syndicator for the right to air a movie or episodes of a television series for a limited number of times. Essentially, a syndicator has an intermediary (or distributor) role: having purchased movies or programs from a studio (the producer), the syndicator then makes money by distributing that material to stations or cable networks.

While syndication has been a boon to broadcast stations needing inexpensive programs to fill out their schedules, basic cable networks have probably needed it even more. With the tremendous cost of renting space on a satellite transponder, along with the uplinking equipment needed to send programming to that transponder, start-up cable networks typically have not had a lot of money left over to produce new programs. Therefore they have relied on the fact that the population has an apparently endless appetite for reruns (both movies and TV shows) and many people are willing to subscribe to cable just to watch more of them. In fact, it is said that Ted Turner successfully launched his TBS superstation in 1976 in part by counter-programming the evening news with *Star Trek* reruns.

How Cable and DBS Companies Make Money

While premium cable and satellite networks such as HBO, Showtime, and Cinemax can offer schedules comprising exclusively recent movies, original TV series, and major sports events, this comes at a high cost to cable and satellite subscribers – who can pay as much as $30 extra per network each month. There are three main reasons why basic cable networks are so much less expensive than premium networks. First, nearly all of them are partially supported by advertising. Those that are not (such as C-SPAN) have alternative revenue sources.

Second, most carry per-sub fees that are folded into the total subscription cost (a notable exception being home shopping channels, which of course have another revenue source). Third, basic networks are bundled, as discussed above, meaning subscribers cannot select individual networks to receive.

At another level, the cable industry has grown wealthier as it has also grown more and more concentrated over time. Through the 1950s US cable was dominated by small, locally owned systems. However, during the 1960s a trend was beginning in which large corporations began to buy out the small systems (often when their local owners wished to retire or move on to other business ventures). These corporations are known as multiple-system operators, or MSOs. The logic driving MSOs is that the more systems one company owns, the more it enjoys economies of scale. In other words, it costs much less for a single company to operate multiple cable systems from one central headquarters than it would cost, cumulatively, for a number of individual systems to be operated separately. Today, two corporations, Time Warner and Comcast, control the vast majority of local cable systems in the United States. Additionally, cable MSOs have tended to be owned by other media or telecommunications conglomerates, thereby contributing to and benefiting from the synergy generated within those corporations (see Box 1.1).

Today's US direct broadcast satellite (DBS) industry is similarly concentrated, with two companies – DirecTV (owned by the News Corporation media conglomerate) and Echostar – providing all of the equipment and distributing nearly all of the programming. With DBS, however, the concentration has been more an effect of the daunting start-up costs than of capital accumulation over time.

Other National Media Scenarios

As discussed above, this book gives some brief attention to select multichannel television scenarios outside the United States. In order to fully understand the development of US television, it is essential to recognize that television developed differently in other nations. Some countries, particularly in Latin America, developed their television infrastructures according to the US model of commercial sponsorship. In fact, the US broadcast networks and other US-based media

Box 1.1 Time Warner and Media Synergy

The 1989 merger of Time, Inc. (book and magazine publishers and owners of Home Box Office) with Warner Bros. (producers of movies and recorded music) heralded a new era for the entertainment and information industries. Not only was the newly merged Time Warner corporation able to combine its existing businesses; it was in a better financial position to pursue newer media such as cable television, home video, and internet. Time Warner and similar merged media corporations that followed are called media conglomerates.

Media conglomerates are particularly well suited to benefit from the phenomenon of synergy, in which the effect of all of the components working together is more powerful than the cumulative effect of individual components working in isolation from one another (put more simply, "one plus one equals more than two"). Synergy usually results from a combination of windowing and repurposing. In windowing, a single media product such as a movie or television program can be "recycled" indefinitely through multiple distribution or exhibition venues (a movie, for example, appearing in theaters, then on home video, then on cable or DBS, and then on broadcast television). Of course, a media product needs to be produced only once, even though it can be used forever. And because a media conglomerate is likely to own many of the distribution or exhibition outlets that product passes through, the conglomerate has the potential to make money at each stage of the life of a media product.

Repurposing is similar to windowing, except that the recycling occurs not just with the actual media product, but also with the entire theme surrounding that media product. Thus the synergistic circle is broadened to include, for example, books related to a popular movie, television programs that use the movie's characters in new stories, and video games based on the movie. With the rise of media conglomerates, people in the entertainment industries became interested in both the movie itself and its "franchise" – in other words, the ancillary opportunities brought about by windowing and repurposing.

In terms of synergy, it might be said that Time Warner hit the ground running in 1989. That was the year the movie *Batman* was released (directed by Tim Burton and starring Michael Keaton, Jack Nicholson, and Kim Basinger). A glance at the home-video version of the movie's opening credits shows how much synergistic potential the movie had due to the holdings of the new conglomerate: story based on a DC Comics character, soundtrack released on Warner, records, and released for home viewing on Warner Home Video. In addition, the home video includes a commercial for the Warner Bros. catalog (featuring products with Bugs Bunny and his cartoon friends) prior to the start of the credits. Of course this movie has continued to benefit from its original synergy – not least in the form of subsequent Batman movies, each of which has been a franchise in its own right along with building the larger franchise.

Then in 1996, Time Warner merged with Turner Broadcasting, thereby adding several popular cable networks (including CNN, TNT, TBS Superstation, and Cartoon Network) to its already respectable stable. In 2001 it merged again, with America Online (AOL), to add a large array of internet services. These sorts of synergistic mergers and acquisitions will be discussed at length in Chapter 7.

companies were heavily influential in the launch of commercial television broadcasting entities in nations such as Mexico, Brazil, and Venezuela. Though it would hardly be correct to claim that television in these nations has developed similarly to that of the United States in the ensuing years, these nations rank high among the world's exporters of non-English television programming.

Other nations developed broadcasting under direct government control, sometimes as a mouthpiece for communist or totalitarian regimes. The persuasive, propagandistic potential of broadcast media is well known and has been relied upon heavily by those seeking either to gain or maintain political power. Not surprisingly, governments in these nations have faced more challenges as satellites make it easier for their citizens to receive programming from outside.

When nations in the developing world began to sever their colonial ties (political as well as economic), many quickly became totalitarian states. Others, though, sought to build more democratic broadcasting systems – only to end up filling much of their program schedules with inexpensive imported programs, primarily from the United States (and to a lesser extent, Western Europe). This planted the seeds for the cultural imperialism that would only be exacerbated as multichannel television technologies made it ever easier to import and export programs.

More prosperous nations, particularly those in western Europe, the British Commonwealth, and Japan, adopted some sort of "public service" model, in which broadcasting has a great deal of government oversight, but nonetheless has developed with a degree of autonomy from the government. The goal in this model is to make television service as universal and democratic as possible. Common features of the public service model include the following:

- Support through some form of taxation.
- Universal service regardless of population density.
- Efforts to make program selection reflect the ethnic composition of the population.
- A balance of entertainment with educational/informational programs.
- A goal of social improvement through the effective selection and balance of programs.
- Assured compliance with programming mandates through dedicated regulatory bodies.

Interpretations of the public service model vary among the different nations adhering to it. Because the public service type of media system operates independently of the particular government in power, or at least has provisions to reflect more than one political position, it has been somewhat resilient to changes. Today satellite and cable television options exist alongside the traditional broadcast television channels in these nations. However, the newer media tend to conform more to the advertiser-driven model than to the public service model. The newer television options have caused the public service systems to make some compromises regarding their original mandates in order to remain viable, and many analysts doubt that the public service model has much of a future in the era of satellites and broadband communication.

As mentioned above, for many years, the public service model stood as an ideal for nations in the developing world as well. Not surprisingly, more democratic media began to be perceived as both vehicles for and products of more democratic forms of government. Unfortunately, nations desiring this sort of media reform have faced huge economic obstacles that made indigenous programming prohibitive. It was much more feasible to import programs.

Multichannel television technologies have been both a boon and a burden to these nations. On the one hand, they can be an economical way to distribute indigenous programming where it is being produced. This is particularly true of nations with large territories such as Brazil, India, and South Africa. Countries such as these have been successful enough with indigenous production in recent years that they have become programming exporters. On the other hand, multichannel technologies have enabled cultural incursions on a scale previously unimaginable. National governments have attempted, with very little success, to stem the flow of transborder broadcast and satellite signals. The more alluring foreign programming is to people living in poorer countries, the less likely those countries will be to develop successful national television infrastructures.

This sort of globalization is probably inevitable – and not entirely lamentable. Those nations able to produce movies or television programs in any quantities at all now have the potential to have them seen all over the world, perhaps leading to greater cross-cultural understanding. Information products do, after all, flow both ways across national borders. However, historically these flows have been very uneven, with developed Western nations – particularly English-speaking nations, and more particularly the United States – providing a disproportionate amount of the programming. Moreover, by this stage precedents set by wealthier nations for what is popular and lucrative have become deeply entrenched.

Globally, multichannel television has tended to develop according to the American model of advertiser-supported programming, a model in which, critics have claimed, there is little room for taking artistic risks, addressing minority interests, or celebrating indigenous cultural traditions. Yet at the same time, television is crossing national borders more than ever before. Diasporic communities around the world have more access to television programming from their native countries, and the plethora of new cable/satellite channels, especially in the

United States, increasingly fill their schedules with imported programming. It clearly is a different era from the one when cable was invented – both technologically and conceptually – in North America, as will be discussed in Chapter 2.

NOTES

1 A third subscription service, Sky Angel, also exists.
2 Interestingly, cable systems in communities without television stations of their own often carry the signals of network affiliate stations from more than one city. Thus a cable subscriber might be able to watch the same episode of ER on two different channels at 10:00 PM (Eastern and Pacific) and then find that at 11:00 PM those channels are carrying two different local newscasts.

FURTHER READING

Baldwin, Thomas and D. Stevens McVoy, 1988. *Cable Communication*, 2nd ed. Englewood Cliffs, NJ: Prentice-Hall.
Cable history and technology website. Denver, CO: The Cable Center. Available at http://www.cablecenter.org/history/
Parsons, Patrick R. and Robert M. Frieden, 1998. *The Cable and Satellite Television Industries*. Boston: Allyn & Bacon.

Chapter 2

Cable Pre-history and the Community Antenna Pioneers: before 1960

By the late 1940s, practically everyone in the United States knew about television and wanted to watch it, but not everyone could. A situation such as this inevitably creates a role for tinkerers and entrepreneurs – in this case, people capable of designing new technologies in order to overcome both physical and financial obstacles to small-town television service. This was the role taken on by the community antenna television or CATV entrepreneurs who will be discussed in this chapter. Before exploring the rise of CATV, however, we first need to take a look at the history of broadcasting in the United States more generally. As with most media technologies, CATV first came on the scene as a remedy to known deficiencies of an existing medium – in this case, broadcast television, which itself was the heir to other media precedents. And as with most media, CATV first took care of the existing need before going on to develop uses, conventions, and an industry structure of its own. As discussed briefly in Chapter 1, the role for a CATV or cable television in the United States was largely assured by the fact that the commercial funding structure of the media could not ensure universal coverage by broadcast television stations alone. CATV established its presence by filling the gaps in television coverage and, once established, developed its own identity.

The Rise of Commercial Radio and a Regulatory Framework for US Broadcasting

The advent of the telegraph in the mid-nineteenth century changed people's daily lives more rapidly and profoundly than any communication medium before or since, for it changed the speed at which a long-distance message could be sent – from days or weeks to minutes or even seconds. Determining the best ways for the new medium to serve society was a lot for people, including policy-makers, to contend with. There were no technological precedents to provide guidance, even though what was decided upon for the telegraph would help to shape the regulatory course for many future media. As a common carrier (i.e., an entity that transports persons, goods, or messages for compensation), the telegraph (and subsequently the telephone) could well have followed in the path of the US Postal Service and developed under the direct oversight of a federal government agency. While the telegraph initially was under direct control of the government, within its first few years policy-makers chose instead to place it in private hands in the hope that competition would speed its development.

What they did not foresee at this early stage was how readily networked media industries tend toward monopoly or oligopoly. In a monopoly, one company controls an industry; in an oligopoly, a small group of companies dominate an industry. In the case of the telegraph, one company named Western Union quickly developed into a monopoly. While Western Union did not own nearly all of the local telegraph companies in the country, it did control the greatest percentage of lines, necessitating that other, smaller companies rely on it (*and pay it*) for transmission of long-distance messages. Radio – which came on the scene about half a century after the telegraph – was first used as a "wireless telegraph," intended primarily for sending messages to and from ships at sea. This made it a common carrier as well, and this use established it as fitting the regulatory precedent set by its wired precursor.

As radio's inventors patented more and more new technologies, government policy-makers no doubt were pleased to see this vital mode of communication grow increasingly sophisticated in its capabilities. Surely they were satisfied that private industry had been able to move so quickly. Many lives were saved, for example, owing to the fact that ships in distress could call for help using the radio. Shipboard radio's impact on the outcomes of naval battles during World War I cannot be overestimated: not only were lives saved, but military strategy could be coordinated more efficiently as well.

Yet this also made radio's role in national security a concern. Since radio patents were then held by a variety of interests, including at least one company with foreign ownership, after the war, US government policy-makers modified their original position on radio's control and allowed consolidation of most of the existing radio patents within the domain of a single new corporation, the Radio Corporation of America (RCA). RCA was formed in late 1919, largely from the four most prominent radio companies operating under US ownership: General Electric, Westinghouse, AT&T, and United Fruit. Foreign-owned companies were required to sell their US radio interests to the newly consolidated company. As something of a government-sanctioned monopoly in radio technology, RCA would become very powerful.

By 1920, US radio's control by RCA was well established. At this point, though, the medium's most prominent role had begun to shift from a wireless telegraph to a source of entertainment and information programs. It was well understood by then that radio works just as well for sending a message to multiple, anonymous recipients as it does to a single, known recipient. Even while America was still at war, various parties – from teenage hobbyists to entrepreneurs wishing to sell radio receivers to the general public – had been thinking about new uses for the medium. This came to fruition in the 1920s, as radio stations of various sizes and strengths began to offer music and spoken-word programs. At first the programming was sustained by sales of receiving equipment, but before very long the notion of program sponsorship had emerged.[1] During this decade, commercial radio stations were started, and soon after, extensive networks to connect those stations and supply them with high-quality programs produced for a nationwide audience. The first permanent radio network, the National Broadcasting Corporation (NBC), was formed by RCA in 1926. The second was the Columbia Broadcasting System (CBS) in 1928.

This was a time in broadcast history when federal policy-makers were contending with the fact that the terms of private control originally established for the telegraph (and which had seemed to work reasonably well for both wireless telegraph-style radio and the telephone as well) presented some complications when applied to broadcast media. Among other things, there turned out to be a limit to how many people or companies could broadcast at the same time without, in effect, canceling out one another's messages. Furthermore, some broadcasters seemed to be taking up valuable frequencies on the finite electromagnetic spectrum merely to hear themselves talk – instead of providing any entertainment or useful information to the public. The situation has been characterized as "chaos."

Between 1922 and 1925, four radio conferences were called by US Secretary of Commerce Herbert Hoover to determine how best to assign broadcasting frequencies so as to eliminate the disorder resulting from overlapping signals and establish some standards for who would be permitted to broadcast and under what terms. The main challenge was to adhere as closely as possible to the First Amendment's free-speech provisions, while also addressing the fact that only a limited number of parties could actually occupy frequencies on the electromagnetic spectrum at the same time. The conferences culminated in the 1927 Radio Act, which created the Federal Radio Commission (FRC), a body with the authority to grant broadcasting licenses to private parties. The Act set standards for assigning frequencies, but its only provision regarding broadcast content once those frequencies had been assigned was that the license assignee act responsibly. The provisions were subsequently adopted into the 1934 Communications Act, which replaced the FRC with the Federal Communications Commission, a government agency that has remained the regulatory authority for electronic media in the United States ever since. The 1934 Communications Act is laid out in fairly broad terms, with its main purpose being to define jurisdictional authority for the FCC in order that it might most effectively serve the "public convenience, interest, or necessity" (a phrase more commonly referred to as the "public interest" mandate).

The provisions of the 1934 Act were written with broadcast television in mind in addition to radio, as television clearly was on the horizon by that point. Thus television was to develop under the same permissive programming regulations as radio. Licensing of stations would adhere to the radio precedent. And since television technology was mostly under the control of the same companies that already dominated broadcast radio, it should not be surprising that television also adopted their programming model. Television broadcasting has been predominantly commercially funded ever since. While provisions for educational television were discussed prior to the passage of both acts, relatively little resulted from those discussions.[2]

The Rise of Broadcast Television in the United States

Experiments with television-type technologies had begun in the late nineteenth century. However it was not until decades later, when broadcast radio interests became involved, that the technology and concept we know as television today began to take shape in the United States The owners of NBC and CBS, two of the three major radio networks, believed strongly that audiences were ready for an audiovisual broadcast medium. Radio by this point had developed highly popular program genres such as the soap opera and variety show, and producers of these programs had become adept at using dialogue and sound effects to compensate for the lack of visual cues. But as audience tastes grew more sophisticated, there was a sense that more was needed. Plus, everyone knew how popular movies had become by that point – the 1920s through the 1950s being known as Hollywood's "heyday" or "golden age."

Throughout the late 1920s and 1930s, RCA (NBC's corporate parent), CBS, and a third company, Alan B. Dumont Laboratories, all employed inventors and engineers trying to figure out the best system for television. There were competing patents (including some bitter disputes), proprietary technologies, and some outlandish public demonstrations (including broadcasts from an airplane). But the most extravagant and widely viewed demonstration of television in the United States was by RCA/NBC at the 1939 World's Fair in New York City. Even though what people saw (or simply heard about) in New York would be perceived as a very low quality picture by today's standards, television was very well received. The RCA/NBC system would be the television technology approved by the FCC and adopted by the industry. And CBS would stay competitive using this system, even while maintaining that its own was superior. The comparatively short-lived Dumont network also would adopt RCA/NBC's system.

At this stage, though, very few people were actually getting to *watch* television – other than in the public demonstrations. For one thing, television sets were expensive and scarce. Also, while regulators were busy determining where in the electromagnetic spectrum to place television, how to allocate frequencies, and how many frequencies to allocate, there was a war taking place in Europe. No sooner had the standards issues been settled (at least temporarily) than the United States entered World War II. Nearly all developments with television were suspended and its newly allocated frequencies temporarily given over to wartime radio broadcasts. Not until after the war would regular television broadcasts resume and start to reach the general population.

Then television arrived with a vengeance! At the end of the war, twenty-three television stations were ready to resume broadcasts, having been licensed just prior to the declaration of war. The FCC granted 24 more licenses in 1946. And by 1948 there were approximately 100 television stations licensed and operating in the United States There were four television networks at that point as well: NBC, CBS, DuMont, and ABC. Before the war ABC had been a second and smaller NBC network (NBC-Blue as opposed to NBC-Red, which is the NBC we know today).[3] And sales of sets were booming – even at prices that are staggering when converted into current dollar amounts (the equivalent of about $3,000 for a very basic black and white set).

Perhaps even more revealing of television's growing popularity than either the number of stations or the number of sets being sold, though, was the prime-time television schedule. In 1946, there were evident gaps in each of the networks' evening program line-ups. Two years later, virtually every slot was taken. Americans, ravenous for television fare, were starting to watch such eclectic offerings as *Charade Quiz*, *Fashion Story*, *Youth on the March*, and *Roller Derby*. They were also starting to watch such "classics" as *Kraft Television Theatre* and *Howdy Doody*. But in 1948, just at the point when television was on nearly everyone's mind whether they had actually seen it or

Figure 2.1 By the late 1940s, television sets were becoming standard furniture in US living rooms and dens [Getty Images]

not, the FCC forced another lag period for the new medium – not because of war this time, but because of what was perceived as regulatory confusion.

The freeze went into effect to resolve some important issues – the main one being spectrum allocation to ensure equal availability of television across the nation.[4] Federal policy-makers had long cherished (at least in theory) a notion of "local service" in broadcast media, but this was not working out as well in television as it had in radio. One problem is that television broadcasts consume more bandwidth, and twelve VHF channels were simply not enough to avoid the overlap and signal interference that occurs with cities in relatively close proximity. So part of the Commission's goal during this period was to decide how to assign frequencies in the 70-channel UHF band to supplement VHF. This all took time, for as broadcast historians

Christopher H. Sterling and John Michael Kittross point out, "Having incorporated virtually all its television problems into one omnibus hearing docket, the FCC was not inclined to loosen the leash on new television stations until it had made decisions on *all* problems" (329).

The freeze finally did end, though, with the FCC's issuance of its Sixth Report and Order on April 14, 1952. This document spelled out a scheme for "intermixture" of VHF and UHF stations within individual markets to allow for more stations per market without signal interference. But this almost instantly created an economic problem for UHF stations, whose signal quality was inferior to that of VHF stations. The problem was exacerbated by the fact that most television receivers at the time were equipped only to receive VHF signals. And even though UHF stations licensed in smaller markets seldom faced competition from VHF, those stations – because of the smaller population, their often remote locations, and the less extensive reach of their signals – would face financial difficulty.

A major challenge to the FCC's goal of providing near-universal television service has been that as long as television broadcasting continues to rely almost exclusively on commercial support, it will never thrive in areas of relatively sparse population. There simply is not the critical mass of potential viewers to compel businesses to produce and buy time for advertising. Moreover, even if a smaller or more isolated community can sustain a single station, this pales in comparison with the multiple stations available in larger urban areas. So even where local service *has* been available, it has not necessarily meant *equal* service. Thus it should be apparent throughout this book that the FCC's inability or unwillingness to address this issue sufficiently helped the CATV industry to thrive, even after the freeze ended and many more stations were licensed.

The first community antennas in the United States appeared during the freeze period, the earliest claims citing fall of 1948 as their launch dates. Given the heavy publicity television had been receiving, combined with the fact that only about half the nation's population had regular access to it, it hardly seems coincidental that inventors would emerge to address this need. And given the imperfect license allocation system that emerged out of the freeze, it is not surprising that CATV remained in place long after.

CATV "Firsts"

During the four years of the freeze, community antenna systems rose from a few maverick start-up operations to dozens of local CATV businesses throughout the country. In other words, during this unforeseen period, a distinct CATV industry formed. There are at least four credible claims to the first cable system in the United States. Each has its own merits, as do several other "first" claims in cable history (namely the numerous individual innovations that fine-tuned the basic technology and business strategies and kept the industry moving forward). The main reason for discussing each claim to a first system individually, though, is to create a picture of the geographical and social climate in which cable television became part of the American media landscape. All of cable's inventors were young-to-middle-aged men, all lived in communities where over-the-air reception of television signals was hindered by mountains or distance or both, virtually all had backgrounds in radio technology, and virtually all were trying to sell television sets. Some of them had formal training in electronics through either trade school or military service.

Astoria, Oregon

One man often said to have built the first CATV system is L. E. ("Ed") Parsons of Astoria, Oregon. Parsons had taken an interest in electronics while still a child. As a teen, he made a small income building and repairing radio sets. By the late 1940s, after several other jobs, Parsons was working as a radio broadcaster, having acquired a nearly defunct station in Astoria and brought it back to life. It was in this capacity that he and his wife Grace (formerly a journalist) traveled to a broadcasting convention in Chicago. While there, they were able to watch broadcast television for the first time. As the story goes, Grace was fascinated and pleaded with her husband to bring television to Astoria. She felt that if anyone could make this happen, it would be her talented husband.

He had his opportunity in summer 1948, when KRSC, Channel 5-Seattle, went on the air. After some experimentation, Parsons found that KRSC's signal filtered through the mountains surrounding Astoria in bands or "fingers" of varying strengths. In fact one of the fingers was accessible from the roof of the two-story apartment building where he lived, so he erected an antenna there. He and his wife worked as a team to tune the signal, he on the rooftop adjusting the antenna and she in the living room reporting on signal quality via walkie-talkie. No clear signal was received, however, until he relocated the antenna to the top of the nearby eight-story Astor Hotel, and then connected it to his own building.

It did not take long before the public became curious about Parsons's work, and so both his home and the Astor Hotel became local attractions. In an interview, he recalled that, "A short time later, [the hotel manager] asked me to remove the set because the lobby was so full people couldn't get in to register." Parsons explained that he then approached the owner of a local music and appliance store about hooking his business up to the antenna. The owner was receptive because of the potential customers it would draw, and he even left the set on in the store window overnight (Parsons oral history).

Since Parsons wanted to accommodate the many residents of Astoria who approached him with requests to have the antenna service extended to their homes, he had to develop a relay system. What he came up with was a combination of a coaxial cable network to connect buildings and transmitting antennas for sending the signal across streets. At first he charged subscribers directly for the needed equipment, most of which he had either invented or adapted himself. There was no monthly service charge at first, but by the end of 1951, he had worked out a standard installation fee of $125 and a $3 monthly rental charge.

In the three years following his introduction of CATV to Astoria, Parsons helped build systems throughout the Pacific Northwest. He traveled extensively in his small private airplane, consulting on technology, gaining access to utility poles, and negotiating business contracts with the communities to be served. He also helped prospective CATV operators find the best locations for signal reception. Finally, though, dealing with mental and physical exhaustion, Parsons moved to Fairbanks, Alaska in 1953 and remained there almost exclusively until his death in 1989. During his years in Alaska, he made great innovations in telephone service and trans-arctic communication for airplanes, but retained no affiliation whatsoever with the CATV industry that was growing rapidly in the "lower 48."

Tuckerman, Arkansas

Meanwhile, at almost exactly the same time Parsons was building his CATV system in Astoria, James Y. ("Jimmy") Davidson was doing something similar in Tuckerman, Arkansas. Tuckerman is a very small town 90 miles northwest of Memphis. Davidson actually had been born in Little Rock in 1922, the son of an optometrist and jeweler (who also happened to hold several radio patents). He was orphaned at a young age, however, and eventually wound up living in Tuckerman. As a young adult, Davidson managed the local movie theater and in his spare time ran a radio repair business – a skill he had learned as a teenager. He served the Navy's Signal Corps during World War II and, in a story strikingly similar to those of other community antenna pioneers, returned to Tuckerman to run an appliance store.

Shortly after that, in late 1947, Davidson found out that a television station soon would be starting operations in Memphis. But then he was disappointed that Tuckerman was too far away to receive the station's signal, and so he and an associate set about building a 100-foot antenna tower near his store. During the station's start-up period he watched test transmissions – at first simply test patterns, but eventually a live telecast of a football game between University of Tennessee and University of Mississippi. At the time, Davidson's system had only one subscriber, the local telegraph operator, but he also connected the American Legion hall to the antenna for this special event. A very large crowd was in attendance, and not surprisingly people clamored for CATV service a few months later (January 1, 1949) when the Memphis station began full-time broadcasting.[5]

Like Parsons, Davidson was eager to see other communities in his region gain access to television. Soon after beginning work on the Tuck-erman system, he began to build community antenna systems for nearby towns. He also started a Davco, a supply and consulting business serving those who wished to start community antenna systems of their own. As journalist Tom Southwick explains, "Davco specialized in building a complete cable headend and then flying it intact to the location where it was needed and where it could be installed in a day" (9).

Mahanoy City, Pennsylvania

Meanwhile, in east-central Pennsylvania, in the southernmost portion of that state's anthracite coal mining region, at least two CATV entrepreneurs were getting their start – with several more soon to follow. While there are many factors to consider in accounting for the proliferation of community antenna systems in this one relatively small area, geography cannot be overlooked. The anthracite region is characterized by both steep mountains and narrow valleys. This made an ideal "laboratory" for CATV experimentation: the mountaintops had good signal reception, and it was only a short distance from there to the valley communities needing television service. At the time, everyone in the region was pleased at the prospect of the new community antenna technology, except for bar and restaurant owners in the mountaintop communities, who had been making a nice profit from valley-dwellers wishing to watch television in their establishments. Within the span of about three years, nearly every community in the region had access to CATV.[6]

The first CATV system built in the anthracite region probably was Service Electric Television, owned by John Walson (formerly Walsonavich) of Mahanoy City, Pennsylvania. As is the case with many other former anthracite mining communities, most of the homes and businesses in Mahanoy City are row-style buildings on the sides of the mountains, structures little changed since they were built during the mining boom a century ago. A visitor to Mahanoy City can easily perceive how the architecture and topography were conducive to mountaintop signal reception and subsequent relaying of that signal via a wire network to multiple houses.

Like many other CATV pioneers, Walson had received training in electronics – in his case, in Chicago during the 1930s (his first exposure to the coaxial cable he would eventually use in Mahanoy City). He also had experience with wiring and pole attachments from concurrent work for Pennsylvania Power & Light (PP&L) – a job he continued for a few years after starting his CATV business. During the late 1940s, Walson had the opportunity to take a second job selling appliances, including televisions. He knew he would be able to sell sets to people living in nearby mountaintop communities such as Hazleton and Frackville, even if he could not sell to residents of Mahanoy City. However, this was far from desirable. As it turned

Figure 2.2 Mahanoy City row houses. Row-style houses in Mahanoy City look much the same today as they did in the 1950s [author photo]

out, nearly all his potential customers were reluctant to purchase sets they could not see demonstrated in the store. Thus, he explained, "The only way that I could demonstrate those TV sets was to build a tower site on top of a mountain, and put a building up there" (Walson oral history).

He did just that – and to his surprise, people not only wanted to purchase television sets from him but also wanted him to connect their homes to his mountaintop antenna. Walson began to offer the community antenna service for free with the purchase of a set. Within its first year, he claimed, the system had close to 1,000 subscribers (a good portion of the town at that time). As of 2007, Service Electric Television is one of the largest cable MSOs, serving a large portion of eastern Pennsylvania and western New Jersey.

Unlike most other CATV pioneers, Walson faced competition for his community's CATV business. In 1950, according to Walson's account, the local police chief, A. P. McLaughlin, began serving customers on the side of town not yet covered by Service Electric.[7] He did this successfully for many years, and it was not until 1970 that Walson actually bought out McLaughlin's system. In response to the competition, Walson had devised one of the first *programming* innovations in cable television history: he stacked a number of receiving antennas, one on top of the other, in order to bring in two New York City independent stations (in addition to the three from Philadelphia he was already carrying). Before long, McLaughlin had devised a five-channel system of his own, and Mahanoy City gained attention for having *two* of the most advanced CATV operations in the country.

This gives a clear sense as to how and why CATV innovation spread so quickly and prolifically throughout the southern anthracite region. A business that relies on very large exterior equipment cannot withhold technology secrets for very long. And when dozens of communities in close proximity have demand for the same type of service, it seems fairly certain that similar businesses will cluster in the area. While it was not common for one town to have two competing CATV businesses, it was extremely common for separate CATV businesses to begin in communities only a few miles apart. In the early 1950s, it was only starting to be realized that the joining of neighboring CATV systems would create tremendous economies of scale in terms of centralized business operations and interconnected technology – a precedent that would drive the monopolistic practices of the future cable industry.

It needs to be noted that Walson's claim is somewhat disputed due to the fact that all records documenting the date he started business were destroyed in a 1952 warehouse fire (see Parsons 1996). At the very least, though, Walson's story represents the degree to which early CATV lay outside the interests of the established entertainment industries, much less the general public. Even if he did launch the nation's first CATV system in 1948, it might have been a "tree falling in the forest" situation. Few if any news media could have perceived the significance of what had occurred because there was no context for it. Anyone living in a metropolitan area such as New York or even Philadelphia – the nearest hubs of "big media" reporting – would not have been aware that parts of rural America were unserved by broadcast television and thus had their lives changed when CATV service came to their communities. The most coverage it would have received as early as 1948 might have been a local news article or perhaps a few lines buried in a larger newspaper or trade publication. But it should not be surprising if even this did not happen.

Lansford, Pennsylvania

Panther Valley TV (PVTV), the system begun by Robert J. Tarlton in Lansford, Pennsylvania in 1950, apparently was the first to receive any press coverage. Articles about it appeared in 1951 in *Wall Street Journal, Radio & Television News*, and *Television Digest* – and so one might say that Tarlton was responsible for making the development of CATV known to people outside the anthracite region.[8] Lansford (sixteen miles north-east of Mahanoy City) was an anthracite mining valley town with geography similar to that of Mahanoy City.

Tarlton was born in Lansford in 1914 and raised there. His father had been a boilermaker, employed by the coal mining industry, and who in later adulthood worked in electronics. Tarlton worked with his father as a teenager and young adult, the foundation of his interest in radio and television technology. Eventually father and son opened their own business to sell and repair radios. Then, as was the case frequently in the CATV industry, Tarlton's electronics expertise was honed by military service during World War II. "In the Army, I got to work on transceivers, transmitters, and receivers. I had a lot of experience with those because that was my job overseas. I installed all of the radios in all of the vehicles in the field artillery outfit to which I was assigned. And I maintained them. That gave me valuable experience in transmission and expanded my knowledge of basic electronics" (Tarlton oral history).

It was not long after returning from the war that Tarlton became interested in television reception in Lansford. The nearest station in operation at the time was Philadelphia's CBS affiliate, WKYW. It could be received on mountaintops near Lansford, including in the nearby town of Summit Hill, where Tarlton's uncle lived. He remembers selling "dozens and dozens of sets up there." In a story similar to that of Walson, he recalls that, "We sold television sets in the Summit Hill area. In Lansford, in a valley one half mile below television reception was non-existent. Most frustrating. And that was the beginning of efforts to get television down in this valley" (Tarlton oral history).

Tarlton also remembered that during the late 1940s, he and other Lansford residents began placing antennas part-way up the mountain and then connecting them to television sets with twin-lead cable. Word of Tarlton's extraordinary capability in this area got around. At one point he was approached by a prominent local acquaintance asking him to rig up an antenna at his home to impress a visitor.[9] Tarlton was able to do this, probably in 1948. However, unlike some of his contemporaries, Tarlton did not wish to claim a "first." Although his makeshift technology was popular, he considered it an imperfect set-up at best, since under certain atmospheric conditions the signal was poor or non-existent. As Tarlton explained, "That stuff won't last, you know. And inside of six months to a year they'll no longer have television or it'll be so poor they'll come back to us and demand their money back. That struck in my mind as being an important consideration" (Tarlton oral history). So unlike the system Walson claimed as a first, PVTV was a professionally run CATV business from the start, contributing to the ambiguity as to cable's "true inventor."

Tarlton managed to find local investors for PVTV early on – mostly individuals working in appliance stores that had been trying unsuccessfully to sell television sets. Tarlton and each of four other backers put in $500, $20,000 more was borrowed from local banks, and the rest of the start-up costs came from the $100 installation fees charged subscribers. Of course, this sort of bank loan was extraordinary for the time period and owes almost everything to a combination of the investors' local reputations and the promise of television. Tarlton explained, "You should understand that television was magic in those days. I showed [one banker] what it looked like and explained what we were going to do. There was no problem whatsoever. He said, 'The Board will pass this without any trouble.' But there again it was the fact that all our personal signatures were on the note" (Tarlton oral history).

Within a few years, CATV systems had sprung up in other anthracite region communities, including Pottsville, Schuylkill Haven, Tamaqua, Shenandoah, St Clair, Mount Carmel, and Shamokin (among a number of others). By 1953, CATV had also spread outside the region and into neighboring states. CATV systems near Astoria, Oregon and Tuckerman, Arkansas had proliferated as well. And, in the next few years, as the industry grew more cohesive and formed

Table 2.1 Number of US cable systems and subscribers, 1952–60

Year	Systems (no.)	Subscribers (no.)
1952	70	14,000
1953	150	30,000
1954	300	65,000
1955	400	150,000
1956	450	300,000
1957	500	350,000
1958	525	450,000
1959	560	550,000
1960	640	650,000

Source: *Television & Cable Factbook.*

its own trade association, the number of systems grew dramatically nationwide.

Equipping the New Industry

When PVTV officially went into business in spring of 1950, the system was built using coaxial cable and components manufactured by Jerrold Electronics of Philadelphia and designed specifically for CATV use. Tarlton had approached Jerrold's owner, Milton Jerrold Shapp, upon learning that Jerrold was in the business of supplying "master antenna" systems that served hotels and apartment houses in television-served urban areas. Shapp quickly perceived the potential for equipping CATV systems once Tarlton, and soon other CATV pioneers, began purchasing components for their systems. In fact Lansford's CATV system became the showcase as well as the testing ground for Jerrold components. In the course of a few years, Jerrold became the largest supplier of components to the growing CATV industry. While there were other electronics firms serving the industry at the time, Jerrold is notable for its dedication to CATV specifically. At a time when dozens of new local systems were sprouting up every year, this proved a lucrative endeavor.

Box 2.1 The Extraordinary Life of Milton Jerrold Shapp

Milton Jerrold Shapp was born in 1912 in Cleveland, Ohio as Milton Jerrold Shapiro. The family was Jewish, and as a young adult, Milton would change his name from Shapiro to Shapp in an attempt to avoid the anti-Semitism rampant at the time. Milton Shapiro was the son of politically active parents: a staunchly Republican father, Aaron Shapiro, and a Democrat suffragette mother, Eva Smelsey Shapiro. His childhood exposure to politics would prove valuable to him later in life.

Like most CATV pioneers, Shapp developed a childhood interest in electronics. He recalls that:

> Early on when I was about eight or nine or ten years old, I got involved in ham radio. I was not a ham but there was a very close friend of mine, or I should say our family, who had a remarkable ham radio set and lived on Lake Shore Drive outside of Cleveland. About once a month I'd go out there and spend the day. I was a kid at that time maybe 11, 12 years old. Oh glory, and that got me, of course, involved in electronics and as I grew I got more deeply involved in radio and so when I graduated from high school, I decided I was going to study electrical engineering. (Shapp oral history)

Shapp thus went on to earn a degree in electrical engineering in 1933 from what is now Case Western Reserve University. After college, he worked as a salesman for Radiart, a manufacturer of radio components. Upon becoming a regional sales manager, he relocated to Philadelphia.

Shapp served in the US Signal Corps during World War II, and became interested in television technology upon his return. While trying to start an electronics business in Philadelphia, he heard about a graduate engineering student at the US Naval Academy who had designed a one-tube "gain-box" or booster that attached to a home television antenna, eliminating signal interference. Shapp realized that this would be a boon for television reception in urban areas.

This booster led to the development of a master antenna system called MUL-TV. Master antennas are receiving antennas placed atop urban buildings (department stores, hotels, and apartment buildings) and connected by cable to various locations within the buildings. Several cable pioneers have related experiences of traveling to cities on business, watching television in hotel rooms, and querying hotel staff on how this sort of technology worked.

Master antennas were considered state-of-the art technology when Robert Tarlton was starting his CATV business in Lansford. Tarlton was adapting the Jerrold components for a new use, however, and it did not take long for Shapp to become curious. Once he found out what was going on, he quickly made Lansford a test site for Jerrold's CATV components. Shapp eventually hired Tarlton to help supervise the building of other systems. Since formal training in community antenna technology was nonexistent at the time, someone like Tarlton, who had tinkered with existing technology to make a functional CATV business, was undoubtedly the best hire Shapp could have made.

Tarlton was especially helpful with the CATV "service agreements," under which Jerrold would design, build, and maintain a system for people wishing to start CATV in their communities but who lacked the necessary technical expertise. Shapp had even convinced some New York City banking houses to help back this venture. This was precedent-setting, and a great many local CATV systems got started in this way. Jerrold, of course, made money, so some might say it was a mutually beneficial agreement.

Unfortunately, not everyone felt this way. In 1957, the US Justice Department launched an antitrust suit to end what it felt was too much control of an industry by one vertically integrated company. Only a decade earlier the Justice Department had forced Hollywood movie studios to sell off their theater chains for similar reasons. The fledgling CATV industry had learned an important lesson. When the suit was eventually settled in 1960, Jerrold was forced to discontinue the service agreement practice.

Shapp was already beginning to sell off his personal interest in Jerrold by this point anyway, and starting a second career in politics. He made unsuccessful bids for both US Senator and Pennsylvania Governor during the 1960s, but was more successful in the next decade. He served as Democratic Pennsylvania Governor from 1971 to 1979. During those years, Shapp drew national attention for consumer advocacy policies and programs to serve elderly and disabled people. He was also recognized for the extensive relief efforts he undertook in 1972 when the powerful Hurricane Agnes struck Pennsylvania.

After leaving office, Shapp went into retirement and attempted to complete an autobiography. In November 1994, however, he succumbed to the devastating effects of Alzheimer's disease and never had the chance to finish telling his own fascinating life story.

But Jerrold was not merely in the business of selling components; it was also heavily involved in designing and building systems. Shapp and his associates supplied components to CATV operators, kept careful track of how they were using those components as well as any changes they made to them, and continually supplied them with more sophisticated equipment. Throughout the 1950s, this was all part of what Jerrold called its "service agreements." Shapp was an extraordinarily perceptive businessman. In the course of working with PVTV and other very early systems, he figured out that most individuals wishing to start CATV systems in their communities would lack Tarlton's technical ability, his local financing connections, or both.

The Formation of an Industry

Another thing that makes the anthracite region stand out in cable history is that one of its communities, the city of Pottsville, Pennsylvania, was the site of the earliest meetings of the National Community Television Association (NCTA), the organization that would grow into the National Cable Television Association and later the National Cable and Telecommunications Association. In all of its incarnations, this has been the trade organization at the center of cable's political, economic, technological, and programming developments over the years. The NCTA had its first annual convention (following several organizational meetings) in Pottsville's luxurious Necho Allen Hotel in January 1952. There were over 30 attendees present, representing around half of the CATV systems in operation at the time, plus a few

Figure 2.3   Milton Jerrold Shapp [The Cable Center]

other individuals interested in CATV. It was clear that by this point, CATV had spread beyond the anthracite region, even beyond Pennsylvania. Among those at the convention were operators from Maryland, Ohio, and New Hampshire – as well as communities throughout the state of Pennsylvania (Community TV group).

The organizer of the NCTA, and its first president, was Martin Malarkey, who had founded the Pottsville CATV system, Trans-Video, in the previous year and was well known to other operators in the area. He decided to organize the trade association in response to some common problems he had heard about. As he explained:

Number one, we had problems with the telephone company. They were rather reluctant, after the initial flurry of granted permissions, to let cable on their poles. They were becoming more and more reluctant to permit attachment to their poles. Number two, the Internal Revenue Service was insisting that we collect a tax called an excise tax from the subscribers. Number three, the [Federal Communications] Commission was beginning – at the urging of the television broadcasters – to evidence more and more curiosity and beginning to ask questions about what we were doing and where we were going. (Malarkey oral history)

The tax issue would not be resolved for several more years. The pole attachments would take even longer. And the FCC "curiosity" Malarkey referred to would follow the industry for all of its existence – as is the role of a federal regulatory agency – continually presenting new challenges. Those gathered realized they would need various sorts of professional assistance. Thus they set about hiring legal counsel and setting codes of business conduct for their industry.

Regulatory Challenges and Legal Experts

The NCTA's first General Counsel was attorney E. Stratford Smith, who had attended the organizational meetings out of fascination with the new technology and those who were developing it. Smith had been working for the FCC's Common Carrier Bureau before becoming affiliated with the NCTA. This is where he first learned about CATV. Because CATV did not fit one of the FCC's existing regulatory categories, common carrier or broadcast (see box), its existence presented something of a regulatory dilemma. So Smith was dispatched to investigate its functions. He reported back that CATV did, indeed qualify as a common carrier – a position he subsequently tried to rescind on more than one occasion. Once he became more aware of the range of CATV's capabilities, he felt that regulating it as a common carrier would be too restrictive – limiting it to a simple retransmission function rather than allowing it to develop its own programming (Smith oral history and personal communications).

Box 2.2   The FCC and Multichannel Television

The Federal Communications Commission (FCC) was created by the 1934 Communications Act. As a federal government *agency*, the FCC is part of the executive branch of the government, along with the president and vice-president, the cabinet, state governors, local mayors, and the police. Other federal agencies include the Federal Bureau of Investigation (FBI), the Internal Revenue Service (IRS), and the Food and Drug Administration (FDA). Unlike the legislative branch (US Congress, state legislatures, city councils) that make laws, the job of the executive branch is to *enforce* laws. In the case of the federal agencies this involves making rules or regulations that refine laws and make them applicable to specific situations. For example, the main provision of the 1934 Communications Act is that the media must act "in the public interest, convenience, or necessity." It is the FCC's job to specify what this means exactly by doing things such as setting compatibility standards for consumer electronics components, setting standards and soliciting consumer feedback for a station's license renewal, and ensuring the provision of telecommunications services to rural residents.

At the time the first CATV systems were being built, one FCC bureau was the Common Carrier Bureau. Its jurisdiction included telegraph and telephone: wired, point-to-point technologies used to transmit messages regardless of their content. The Commission's other main regulatory jurisdiction, broadcast media, included radio and television. These media were regulated differently from common carriers. First, they relied on the electromagnetic spectrum (i.e., the "airwaves") for transmission. With the spectrum's limited number of frequencies, it was the FCC's job to ensure that those licensed to use it were doing their best with regard to "the public interest." Second, because broadcast messages were, by nature, available indiscriminately to many different listeners or viewers, the FCC needed to set in place measures to ensure responsible programming.

CATV, of course, fit neither of these FCC jurisdictions, because it was a wired medium, like the common carriers, intended to transmit the messages of a broadcast medium. Therefore it presented a regulatory dilemma for the FCC.

Today's FCC is structured to accommodate cable and DBS, along with broadcast television and radio, within the Media Bureau. Other new media have also brought about some restructuring. For example, the former Common Carrier Bureau has been replaced by the Wireline Competition Bureau (wired telephones) and the Wireless Telecommunications Bureau (cellphones, pagers, etc.).

The tax problem

Smith offered his legal services to the NCTA in 1952, by which point he had entered private practice in Washington, DC. At the time Smith began working with the CATV industry, the revenue tax issue was still pending. Subscribers in most communities were paying taxes equal to eight percent of their $2–4 monthly fees and their $100–200 installation charges. The justification for the tax was that CATV was a "leased wire service" similar to the news wire services, stock quote services, and the AT&T long line services that broadcast networks used to deliver programs to their affiliate stations – all taxable under the Internal Revenue Code. The CATV industry believed otherwise, feeling that since its subscribers had no control over the content delivered to them, their service was of a different nature. They challenged the tax by instituting litigation involving Meadville Master Antenna (MMA) of Meadville, Pennsylvania and one of its subscribers, Gust Pahoulis. MMA's owners, the father-daughter team of George and Yolanda Barco, were also attorneys. They tried the case, *Pahoulis v. United States*, in Pennsylvania district court and lost. Meanwhile a similar case, *Lilly v. United States*, had been lost in West Virginia district court. The Barcos then appealed *Lilly* and won, overturning it and then, by judicial precedent, *Pahoulis* as well. The government ultimately had to refund over twenty million dollars to CATV subscribers. The taxation issue was settled, at least for a while (Smith).

The microwave problem

An even thornier issue arising for the industry during the 1950s had to do with the use of microwave relays. By the middle of the decade, several CATV systems in western states had started using this technology, since some communities were located too far from broadcast stations for simple community antennas to be of much use. Also, once channel capacity on most systems had grown from three to five or more, some operators (including Walson) had begun to think of microwave as a way to enhance the channel selections they offered subscribers – a smart marketing strategy that eventually would make CATV service desirable even in smaller broadcast markets. The technology prefigured a future in which CATV/cable would not simply retransmit broadcast television, but would present serious competition for it.

Small-market television stations (in places such as Cheyenne, Wyoming) expressed concern that CATV would put them out of business by using microwave relays to import the signals of better-funded stations from larger cities. Though station owners were troubled by CATV generally, they were smart to couch their concerns in the form of complaints about CATV's use of microwave relays specifically. CATV by itself actually did not fit either of the FCC's two categories of oversight with regard to electronic media: broadcast and common carrier. Because CATV's transmission technologies and the frequencies it used were not available to anyone wishing to pay for the use of them in the way that those of the telegraph and telephone were (and still are), it could not be considered a common carrier. Microwave relay businesses were, however, considered common carriers. The FCC was still holding to its flawed belief in universal broadcasting service, and so when challenges to CATV were presented as complaints about microwave operations specifically, in other words falling with their known jurisdiction, the commissioners paid careful attention.

The first microwave-based challenge to the CATV industry occurred in 1951, when J. E. Belknap and Associates, a microwave outfit in Poplar Bluff, Missouri, filed an application with the FCC to transmit the signal of a Memphis station to its own CATV system as well as others in the area. The Memphis station challenged on the grounds that this use of its signal was illegal – easily seen as a sort of theft since the station would make no money from the retransmission. Little came to pass as a result of the Belknap situation. While the Commission felt qualified to regulate microwave relays as common carriers, it was not yet ready to make a determination about CATV systems themselves. Belknap's application was finally approved in 1954 (but they never actually launched the venture).

The second microwave case was *Frontier Broadcasting v. Collier*, filed with the FCC in 1956, in which thirteen television stations in western states (one of which was called Frontier Broadcasting) filed suit against 288 CATV operators in 36 states. These were the small-market broadcasters mentioned above. They already felt under siege from the booster and translator stations that had been put up by residents wishing to watch the more popular big-city stations, and they saw CATV, a growing industry with multichannel capacity, as an even bigger threat.[10] The station owners saw CATV as a drain on their revenues, feeding advertising dollars to the larger imported stations instead of to them. As *Broadcasting* magazine reported in 1958:

> William C. Grove, KFBC-TV Cheyenne, Wyo. – Has spent over $500,000 in building Cheyenne and Scottsbluff, Neb., stations. Catv systems in "heart" of stations' coverage area – in Laramie, Rawlins, and Pine Bluffs, all Wyo.; Scottsbluff, Alliance, Kimball, and Sidney, Neb., and Sterling, Colo. All being fed Denver signals via microwave relay systems connected to catv systems. Advertisers know of this "bonus" coverage and aren't interested in buying Cheyenne and Scottsbluff time.

When the FCC finally made its decision on *Frontier* in 1958, it essentially bypassed the microwave issue and upheld the position that CATV itself was not a common carrier – and thus was not subject to its regulatory authority.

Legislation: a false start

More was to come, however, as the western broadcasters were quite dissatisfied with this outcome. While virtually no definitive policy-making was done during the 1950s, the later part of the decade saw a great deal of preparation for the *extensive* policy-making to take place during the 1960s. The various complaints left unresolved by the FCC, along with the agency's own questions regarding its authority over CATV, eventually led to an investigation by Congress. In December 1958, a Senate committee issued a report based on testimony from television broadcasters, CATV operators, and the FCC. The report, known as the "Cox Report" (since it was authored by the committee's counsel Kenneth Cox), generally opposed the FCC's position on CATV up to that point, affirming the western broadcasters' economic concerns. It recommended that the regulatory agency begin asserting more authority over the new medium.

The FCC replied to the report in a 1959 Report and Order, maintaining its position in *Frontier*: although it did recommend that CATV systems work out some sort of consent and compensation arrangement with the stations whose signals they carried, the Commission maintained that it held no regulatory authority over CATV. At this point Congress, facing considerable pressure from the western broadcasters in particular, began to draft legislation on CATV. The bill that emerged, S. 2653, represented an attempt to merge the stated interests of both television broadcasters and CATV operators: mandatory FCC licensing of CATV systems, mandatory CATV carriage of local television signals (though without a provision requiring compensation to the broadcasters for this), and rules to prevent duplication of programming on different cable channels (as would occur when two or more network affiliate stations were carried by the system).

Broadcasters were surprisingly accepting of the bill, though as Patrick R. Parsons and James M. Frieden point out, "they got much of what they wanted, and what they didn't get – carriage consent – they were pursuing in the courts" (39). Debate on the bill proved extremely divisive for the CATV industry, however. While Smith and some others believed legislation affecting CATV was inevitable, and saw S. 2653 as a reasonable merging of their industry's interests and those of broadcasters, others (led by Shapp) waged a massive – and ultimately successful – campaign to oppose the compromises. The bill was defeated by one vote and subsequently died in committee. Smith continues to express regret about the turn of events (see Parsons and Frieden).[11]

The debate surely heralded the end to the relatively quiet regulatory climate that CATV had enjoyed throughout the 1950s. We can never know exactly what might have transpired for the industry had the bill passed, but it is likely the 1960s would have been more favorable for CATV. As will be discussed in the next chapter, though, the 1960s proved to be one of the periods in which CATV/cable was subject to the heaviest regulation.

Pay-television

When testifying before Congress, one of the appeals made by the western television broadcasters was to save "free TV." Here they were referring to an ongoing debate about the legal status of a form of television known variously as "pay-television," "subscription television," or "toll television." These referred to forms of television supported by direct viewer payments, not advertising. The systems experimented with during the 1950s and 1960s were the precursors to today's premium cable channels – Home Box Office, Showtime, Cinemax, The Movie Channel, and others. Though it would be more than a decade before the two non-broadcast forms of television would merge, one cannot truly understand the development of modern cable television without understanding the development of both of its precursor forms, CATV and pay-television.

There were three experimental pay-television systems during the 1950s, two of which were very directly CATV-related. Two systems used broadcast signals to transmit scrambled programs to subscribers' homes: Zenith Radio Corporation's Phonevision and Skiatron Television and Electronics Corporation's Subscriber-vision. Both Phonevision and Subscriber-Vision conducted trials during the early 1950s, but ultimately were denied permanent operating permits. Since these systems used the airwaves for program transmission, they fell under the FCC's regulatory authority.

A third system, International Telemeter (partially owned by Paramount Pictures), used wire for its program transmission – ensuring, of course, that its development would be watched carefully by the ambitious CATV industry forming on the east coast. In fact, International Telemeter launched a combination pay-television and community antenna operation in Palm Springs, California in 1951. The pay programs, purchased through a set-top coin box, consisted of "chains" of movies being shown in local theaters (all Paramount productions, so copyright was not an issue) and sports events not carried by local stations.[12] Los Angeles broadcast television signals were provided free of charge.

The Palm Springs system lasted until 1955, by which point it had begun to run out of program sources. The company remained active, however, and actively promoted the compatibility of its system with

CATV. In fact, the CATV industry grew more interested in merged ventures with pay-television throughout the 1950s. Jerrold Electronics expressed particular interest since so many of the components it manufactured could be adapted to a wired system such as International Telemeter. It is not surprising, then, that Jerrold made several efforts to implement pay-television options on existing CATV systems and finally helped found an operation called Telemovies in Bartlesville, Oklahoma in 1957.

Telemovies was owned and operated by Vumore Video, a subsidiary of Video Independent Theatres of Oklahoma City. Vumore's president, Henry Griffing, was active in the NCTA and at the time of Telemovies' debut was operating CATV systems in five Oklahoma communities (not including Bartlesville, whose residents received clear off-air signals from Tulsa). It was a flat-rate system, similar to today's premium cable channels. Like International Telemeter, Telemovies offered its subscribers movies chained from local theaters – in this case a beneficial relationship with the exhibitor rather than the producer/distributor. The logic that guided this practice for Vumore and Telemovies was that the system would draw an audience of people who would not bother seeing a movie if it meant a trip to the local theater. Of course this same logic would eventually drive not only premium cable channels, but also the home video industry.

But Telemovies would be long gone before these things ever came to pass; in fact the system only lasted until early 1958 – less than half a year. There were several factors probably accounting for its demise, including increased movie showings on Tulsa broadcast stations and subscribers' wish for a pay-per-view system such as the one International Telemeter and others had been promising. A last-ditch effort to save Telemovies by adding community antenna service and drastically reducing subscription rates only worsened the situation. It had lost a lot of money already. Griffing himself died a year later in a plane crash along with his entire family, so he was not able to pursue the further development of either pay-television systems or CATV.

Pay-television would go through another promising period in its own right during the 1960s before finally becoming one with cable television. That decade would also see pay-television's most challenging opposition – from both regulators and the public. During the

Figure 2.4   Promoting the Telemovies system [The Cable Center]

1950s, though, policy-makers' position on pay-television had been much as it was on CATV at the time: in limbo. The Commission did see fit to regulate the broadcast-based systems, Phonevision and Subscriber-Vision, but it generally left the wired systems alone. There was opposition to pay-television on the part of commercial television broadcasters, just as there was to cable. There was also a concern that powerful Hollywood movie studios, having been forced to sell off their theater chains in the United States Justice Department's 1948 Paramount Decree, would use pay-television as a way to regain a hold on movie exhibition.[13]

The International Status of Cable Television in the 1950s

Today cable television sometimes seems synonymous with United States television, so it might be hard to believe that cable-type technologies already were being used in other countries at the time the first community antennas were built in Pennsylvania, Oregon, and Arkansas. According to British and Canadian cable pioneer Ken Easton, some European and British Commonwealth nations began experimenting with rudimentary forms of coaxial-type cable as means of extending their national radio services as early as the 1920s. These were countries with public-service-type broadcasting operations, and so the "relays," as they were called, were used for transmitting the public service channels as well. In the UK, coaxial-type cable was used to help distribute the programming of the British Broadcasting Corporation (BBC) starting in the early 1930s, when the BBC was strictly a radio service. Countries such as Belgium and the Netherlands, in contrast, are heavily "cabled" today because, once television service had become well established throughout Europe, community antennas were desirable means of augmenting the programming available from their own national broadcasters with that imported from neighboring nations.

In Argentina, coaxial cable networks became widespread because the government in power during the 1960s – when broadcast television was coming into its own in that nation – prohibited the use of affiliate stations to distribute programming (no doubt because of the possibility that such stations would add programming of their own selection to replace what was provided by the government) (see Sinclair, 84–85). If other totalitarian regimes were not aware of this use of technology, surely it would have been appealing if they had known.

Community antenna service began in Canada primarily as a means of making United States channels available there. Canada's own national broadcasting policy, in the tradition of Western Europe, provided for near-universal availability of the national radio and television service, the Canadian Broadcasting Corporation (CBC). Of course retransmission of American television channels was not difficult to accomplish since the vast majority of Canada's population live very near the United States border – allowing community antennas to function well in most places, sometimes assisted by microwave relays. In fact, Canadians' desire to watch US channels in addition to the CBC meant that CATV penetration there during the 1950s and 1960s was actually higher than it was in the United States. By 1960, a CATV system in Montréal was serving approximately 14,000 sub-scribers – making it the world's largest at that time and for some years to come.[14] CATV was also prevalent in metro Toronto and western Ontario as well as Vancouver – places where United States signals were hard to pick up with a home antenna but quite easy to pick up with a community antenna (see Easton).

Other national comparisons in cable's development are difficult at this early stage. Many nations were only beginning to develop broad-cast television. It should be pointed out, though, that at whatever stage in history multichannel commercial television has been intro-duced to a nation (in whatever form or forms it has taken), it has had to contend with legal and regulatory precedents as well as consumer expectations set by and for broadcast media, which in most cases were controlled directly or indirectly by the government.

What is important to realize from the examples above is that national policies have had a great deal of influence on what sorts of television-related technologies have been developed, why they have been developed, and who has developed them. In Western Europe and the British Commonwealth, cable television developed ancillary to the established national broadcasting services – to help those services achieve their mission of universal service. The United States, having already set in place a government-sanctioned practice of advertiser-supported broadcasting, created a place for the CATV industry since that was the only means by which a large portion of the population could have access to any television at all.

NOTES

1   AT&T has been credited with pioneering commercially funded radio with its "toll broadcasting" concept, wherein an individual or business could pay for air time to use in any way desired.

2   Radio frequencies were set aside for educational broadcasting in 1941 and for television in 1953, most of which benefited from the advent of the

Corporation for Public Broadcasting in the late 1960s. This new source of public funds for broadcasting enabled the creation of the Public Broadcasting System (PBS) and National Public Radio (NPR). These organizations face a constant battle for the small share of their operating revenues that comes from public funds (the rest coming from individual and corporate donations).

3   Note that all of these networks, except Dumont, were operating simultaneously as radio and television networks at this point. The wartime separation of ABC from NBC had initially affected only the radio network. Television inherited the split. Dumont would go out of business in 1955, after years of struggle, its demise strengthening ABC by allowing it to affiliate with a number of Dumont's former affiliate stations.

4   Standards for color broadcasting were also addressed during this period.

5   Biographical details on Davidson are based on interviews conducted by Journalist Tom Southwick for articles in *Cable World*, a cable industry trade magazine, and included in his book, *Distant Signals: How Cable TV Changed the World of Telecommunications*. Information is also drawn from Archer S. Taylor, *History Between Their Ears: Recollections of Pioneer CATV Engineers*.

6   The central and northern portions of the anthracite region would soon be fairly well served by broadcast stations in Scranton and Wilkes-Barre.

7   Apparently, McLaughlin had become intrigued by Walson's business when he came to disperse crowds gathered at this store to watch television. Phillips, p. 9.

8   For mentions of Tarlton's system in the national press, see Lucas, Gould, and Saada.

9   The acquaintance was David John Stevens, executive secretary of the local branch of the United Mine Workers. The visitor was John L. Lewis, that organization's national president.

10   A booster amplifies a station's signal and rebroadcasts it on the same frequency. A translator does the same thing, only rebroadcasting it on a different frequency.

11   This also draws from personal communications with Smith.

12   Chaining a movie is a process of using a video camera to capture it off the screen. At this stage, it could not have been recorded on videotape (as would later be the case), but instead was transmitted directly to International Telemeter's facility and then out to subscribers.

13   From the 1920s through the 1940s, a group of five Hollywood studios – Paramount, Warner Bros., RKO, MGM, and Twentieth Century-Fox – operated as a vertically integrated oligopoly, controlling both distribution and exhibition of the movies they produced. This made it incredibly difficult for smaller studios and independent producers to get their movies shown. Ironically, today the major movie studios technically are still barred from theater ownership, but are not prevented from being part of media conglomerates

that control multiple other distribution and exhibition outlets such as home video, cable networks, and broadcast networks.

14   Unlike other CATV systems in Canada, the one in Montréal had begun as an outgrowth of Rediffusion, the largest broadcast relay business in the UK. While there proved to be little demand for retransmission of the CBC signals, readily available over the air, Montréalers were clamoring for retransmission of nearby United States signals. See Easton for details on the system.

FURTHER READING

Lockman, Brian and Don Sarvey. *Pioneers of Cable Television*. Jefferson, NC: McFarland, 2005.

Phillips, Mary Alice Mayer. *CATV: A History of Community Antenna Television*. Evanston, IL: Northwestern University Press, 1972.

Sterling, Christopher H. and John Michael Kittross. *Stay Tuned: A History of American Broadcasting*, 3rd ed. Mahwah, NJ: LEA, 2002.

Taylor, Archer S. *History Between Their Ears: Recollections of Pioneer CATV Engineers*. Denver: Cable Center, 2000.

Chapter 3

Regulatory Beginnings: 1960–7

The 1960s probably were the decade which had the greatest impact on the future direction of cable and other forms of multichannel television in the United States. This was the time when the rural CATV industry truly began to be noticed by parties other than the residents who relied on it for television service. These parties ranged from concerned government policy-makers (particularly the FCC, but also various other federal, state, and local bodies) to large entertainment corporations perceiving CATV as a lucrative business to complement their various other holdings (the beginnings of today's synergistic media conglomerates) to idealistic think tanks perceiving CATV as the medium that would remedy a flawed national television system. There is a lot to cover here, and a look at these years should remind us that society's adoption of new technologies almost never proceeds smoothly or predictably.

In fact, this pivotal decade does not even fit thematically into a single chapter, as the last three years, 1968–70, represented such a reversal of events taking place from 1960 to 1967 that they must be discussed in conjunction with developments in the early 1970s. Therefore the present chapter looks in detail at the first two-thirds of the 1960s, which has been described as "regulatory," "restricted," and even "frozen." By 1960, the CATV industry was well aware of its scope and potential; there were 640 local systems operating in that year, serving some 650,000 subscribers (*Television & Cable Factbook*). And because it had developed as such a cohesive industry during its founding years, innovations and business strategies spread quickly. Still, the early 1960s would prove to be an extremely frustrating time for such an ambitious group of individuals and companies. For even while those in the industry were well aware of ways to expand their businesses – for example, by moving into larger markets, merging with pay-television systems, and selecting from among the most desirable broadcast stations to retransmit – these opportunities were kept on hold by fickle and uncertain policy-making efforts.

Multiple-System Operators

By 1960 the CATV industry as a whole had begun to see the bigger picture of how its business functioned economically. A number of CATV pioneers had begun to buy out systems in neighboring communities, creating regional consolidation of ownership. By doing this, they could enjoy greater economies of scale – particularly cost savings from multiple operations being run through a central headquarters and coordinated management structure. Other pioneers had sold out their original systems in order to build elsewhere. In addition to the prospect of serving larger communities or regions, and thereby increasing subscriber numbers, there were more tax benefits in starting a new system than there were in maintaining an existing one.

Of course, the idea of multiple-system ownership on a large scale had been developed first by Milton Shapp and Jerrold Electronics, with their innovative financing arrangements and service agreements. By the start of the new decade, however, Jerrold's influence was somewhat diminished – by a court decision requiring them to end the service agreement practice as well as by Shapp's personal shift from the CATV business to a career in politics. There was also more competition from other component manufacturers. Jerrold would continue to have a major role in CATV and cable for decades to come; the company just did not enjoy the degree of influence it once had.

A major shift for the CATV industry in the early 1960s, not surprisingly, was the large-scale rise of MSOs. Leading the charge on this front was Irving Berlin Kahn, whose career in media was already well established. Kahn had been working for Twentieth Century-Fox both before and after his service in World War II, but left when the company failed to show interest in his public speaker-prompting device, the TelePrompTer. He struck off on his own, marketing the TelePrompTer to politicians, newscasters, movie studios, and a

Figure 3.1   Irving Berlin Kahn [The Cable Center]

variety of other interests. However, Kahn eventually sought out other ventures to boost his company's fortunes. During the mid-1950s he became involved in theater television, a new industry that delivered "big ticket" televised events such as plays, ballets, and especially sporting events to movie theaters using either microwave relays or land lines (as opposed to broadcast television frequencies) for transmission.

TelePrompTer's theater television outfit, Box Office Television, became known in particular for carrying boxing matches. Kahn was quick to perceive the success of theater television as a sign that people were willing to pay directly for individual televised events. The connection between this type of special event program and today's pay-per-view cable events seems obvious, though of course it was unprecedented in the 1950s. Kahn became involved in CATV when future cable magnate Bill Daniels stopped by the office of TelePrompTer's vice president and chief financial officer Monroe ("Monty") Rifkin. Rifkin recalled:

> The receptionist called me and said "Mr. Rifkin, there's a guy out here who wants to speak to whoever's in charge of the company. What shall I tell him?" And I asked where's he from. And she said, "He's from

Denver, Colorado." And I said, "I'm in charge, Irving's gone, so show him in." And this dapper dan name of Bill Daniels, walked into my office, introduced himself, and quickly said "I've been reading all about your company being in closed circuit television, vis a vis the fights. I want to tell you about another facet of closed circuit television." I said "What's that?" He said it's called community antenna television. I said "What's that?" Bill proceeded to tell me how there were probably 400,000 homes in the country at that time who were receiving television via wired systems because they were located behind mountains, or distances from television stations where they couldn't get a good signal. And the only way they could get it was some entrepreneur setting up an antenna, capturing the signal and then wiring it to the customer's homes. Sounded very interesting. (Rifkin oral history)

Rifkin informed Kahn, and so late in 1959, TelePrompTer began to acquire CATV systems. The first was in Silver City, New Mexico, for which the company paid $130,000. Cash flow in Silver City then proved so high that TelePrompTer quickly bought systems in Farmington, New Mexico and Rawlins, Wyoming as well. By the mid-1960s, Tele-PrompTer was one of the largest MSOs in the United States, with 14 systems and over 70,000 subscribers. Most prominent was the TelePrompTer system that served a large portion of Manhattan.[1]

Kahn perceived a huge benefit to operating these systems beyond simply their community antenna revenues: they were the ideal testing grounds for adapting his theater television pay-per-view system for delivery to subscribers' homes – truly the first example of cable pay-per-view. He figured that these CATV systems were remote enough not to draw attention from federal regulators concerned about unauthorized pay-television operations. TelePrompTer also owned enough systems so that the pay-per-view events could be carried efficiently on a closed-circuit network. Kahn was eager to let the CATV industry know about his experiments, and did so, on a grand scale, by appearing before attendees of the 1960 NCTA convention in Miami, not in person but via TelePrompTer's theater television equipment. He spoke about the experiments with closed-circuit pay-per-view and also announced the development of a pay-television delivery system called "Key TV," which, though never actually marketed, reflected prescience in its conception.

So besides being one of the earliest CATV operators to emerge from the established media industries (instead of as a result of small-town

Box 3.1   Bill Daniels

Bill Daniels is legendary among cable television's pioneers. Even among other larger-than-life figures, his story stands out. Daniels was one of the early CATV operators, though hardly the earliest. Born in Greeley, Colorado in 1920, Daniels spent his childhood in various western states. After graduating from New Mexico Military Institute, a prep school and junior college, in 1941, Daniels joined the Navy and trained to be a fighter pilot. He served during both World War II and the Korean War. Those who met Daniels later in his life inevitably recall the thrilling stories about various planes he flew. Unlike other CATV pioneers, however, Daniels never related his success with television technology to his military career. He only went so far as to cite the military discipline as a factor in his success as a businessman.

Daniels was in his early thirties when he saw television for the first time – in a Denver bar. He was working as an oil insurance salesman in Casper, Wyoming at the time. Daniels, having heard about the CATV operations in Pennsylvania, got the idea to build a system in Casper. Casper presented at least one challenge that the Pennsylvania systems did not face, however: it was over 200 miles from the nearest television station in Denver. In a 1986 interview he told the story of going to Pottsville, Pennsylvania to consult with Martin Malarkey (at a cost of $500 per day), contracting with Jerrold Electronics for equipment, and agreeing to pay AT&T $800 per month for use of a microwave relay from Laramie, Wyoming – the nearest point at which the signal of a Denver station could be picked up off the air – to Casper (Daniels oral history). Most people believe this was the first use of a microwave relay by a CATV system. If not the first, it was certainly the longest distance covered by microwave in CATV's early years. Daniels's Casper system demonstrated that CATV could be viable in large western states, where communities often lie at great distances from one another.

Daniels did acquire other CATV systems, but what he is best known for are his innovations in the buying and selling of CATV systems. He began doing this informally in 1955. In 1958, after serving as president of the NCTA, Daniels formally launched his CATV brokerage business, Daniels and Associates, in Denver. In 1964 he bought out his business partners, and by the following year was handling some 80 percent of transactions involving the sale of CATV systems – some very highly valued. By the late 1960s, when cable television was growing by leaps and bounds, Daniels and Associates was known as *the* cable brokerage firm.

Daniels had some other cable-related ventures during his lifetime as well. In particular, he was interested in cable sports channels. For example, in 1985 he launched Prime Ticket Network in partnership with Los Angeles Lakers owner Jerry Buss. Four years later he launched Prime Network for regional sports coverage. In retrospect, the sports programming ventures seem diminished compared to his brokerage. Clearly Daniels will go down in cable history for his early awareness of how much money could be made by acting as a middleman between cable system buyers and sellers, and the financial institutions that would make their transactions possible.

Daniels also was fairly well known outside the cable industry. Although his political career never quite matched that of Milton Shapp, Daniels did run for Governor of Colorado (on the Republican ticket) in 1974. He did not win. But during the later 1980s, he served as Colorado representative to the Republican National Committee. Also, during the later part of his life Daniels was well known for his philanthropic efforts – especially reflected in his contributions to the University of Denver (which now has a school of business named after him and also is affiliated with the Cable Center). Bill Daniels died in early 2000 at the age of 79.

entrepreneurship), Kahn was also one of the first in the CATV industry to pursue the merging of CATV and pay-television. Shapp and Jerrold Electronics had preceded him in both endeavors, but whereas Jerrold had reached something of a plateau by 1960, TelePrompTer was growing by leaps and bounds. Other MSOs were in place by the middle of the decade as well, including the very large H & B American Corporation (27 systems and over 90,000 subscribers by the mid-1960s) and American Cablevision (also with 27 systems and some 70,000 subscribers) (*Television and Cable Factbook*). It is also worth pointing out that other media corporations, including broadcast television interests, were acquiring CATV systems in large numbers. This might seem surprising given broadcasters' vociferous opposition to CATV during the previous decade; however the television landscape as a whole had shifted quite a bit in the intervening years. For one thing, many small broadcasters had already been squeezed out of the market by larger economic forces than CATV alone. Many of the remaining station owners were part of larger corporations that saw CATV more as a lucrative investment opportunity than a threat to their fortunes. And by this stage the FCC had finally started to assert its authority over CATV, perhaps quelling some of the lingering complaints by smaller broadcasters.

FCC Action

Recall that during the 1950s the fledgling CATV industry had felt the constant weight of possible regulations, even though virtually none were actually implemented during that decade. In fact the industry probably could have enjoyed this status even longer had it not been for many operators' adoption of microwave relay service in the late 1950s and early 1960s. As discussed in Chapter 2, microwave relays allow CATV systems to bring in television signals over distances too great for simple community antenna reception. This did not present a problem at first. Rather, microwave technology allowed CATV to operate in very remote communities, especially in western states, where it would not have been able to otherwise. Before long, though, CATV operators throughout the country began to bring in the signals of popular and well-funded stations from large cities, bypassing less popular smaller stations (or at least providing subscribers with the opportunity to choose one station over another). So the broadcasters complaining about CATV were the smaller ones, who feared CATV would bring about their economic demise. After all, why would a broadcast television station complain about a service that only *expanded* its audience?

Without having obtained Congressional authority to regulate cable in 1960, the FCC made a tentative first step toward doing so on its own with the *Carter Mountain* decision in 1962 (32 FCC 459) that was affirmed in appeal (321 F.2d 359), a case involving the license application of a common carrier microwave company, Carter Mountain Transmission Corp., wishing to serve CATV systems in some western states. Without any legal precedent for guidance, the FCC determined that microwave relays did indeed threaten the fortunes of small television stations and therefore denied a license to the microwave applicant, especially since that outfit not only refused to guarantee the local station protection against duplication of its own programs by imported stations, but also refused to require the CATV system being served to carry the local station's signal. As the FCC reasoned, "The fact that no broadcaster has actually gone off the air due to CATV competition at the time the government moved to expand its authority (nor have any since) did not stay the momentum for the expansion of regulatory authority. That some economic impact was merely plausible sufficed as the basis for government concern and government action." The *Carter Mountain* decision clearly fell within the FCC's common carrier jurisdiction, easing the Commission's uncertainty about its role in regulating CATV. But the decision ultimately affected the CATV industry as a whole in such a way as to open the door for a broader regulatory role for the FCC.

By the middle of the decade, the FCC, no longer content with its historically passive position on CATV, began to take even more action – in the form of the 1965 First Report and Order (38 FCC 459) and the 1966 Second Report and Order (2 FCC 2d 725). Both were comprehensive sets of regulations affecting CATV. It is important to note that, as regulations not laws, these still adhered to the general terms set forth in the 1934 Communications Act. Nevertheless, they represented a major and telling reconceptualization of the FCC's role: where a decade earlier the Commission had been unable to justify regulating CATV under either its broadcast or its common carrier division, now it was acknowledging that CATV represented a new category it would need to deal with in its mandate to serve the "public interest."

The First Report and Order was basically a restating of the terms set forth in the *Carter Mountain* decision: requiring carriage of a local station's signal by any microwave-served CATV system operating within that station's predicted grade A contour (see Box 3.2). Additionally, the regulations mandated that local stations be protected against program duplication on imported stations for a period of 15 days before or after a broadcast network broadcast. Since broadcast network programs always have been the most popular programs, they have also represented the most promising source of advertising revenue for stations. The nonduplication provision was meant to assure

---

Box 3.2   Television Stations' Service Contours

A television station's service contours are used to define a cable system's degree of obligation to carry that station's signal, as well as whether or not that station's programming may be duplicated by another station carried by the cable system. Predicted grade A contour is the concentric service area in which a good broadcast picture is computed to be available 90 percent of the time at 70 percent of the reception locations. Predicted grade B contour is the concentric service area in which in which a good broadcast picture is computed to be available 90 percent of the time at 50 percent of the reception locations.

One might assume that the service contours would always be circular in shape. In fact they frequently are not perfect circles. An obvious variable would be the presence of mountains, which could cause service contours to have "amoeba-type" shapes. Other factors affecting contour shapes and sizes include antenna height, channel frequency, ground conductivity, climate zone, and something called the "urban clutter factor." A complex set of formulas known as the Longley-Rice Propagation Model is typically used to calculate stations' grade A and grade B contours.

---

that those revenues went to any local or regional stations first and foremost. Of course, nonduplication was not a concern for the many CATV systems that lay outside any stations' grade A or grade B contours.[2]

It took the FCC less than a year to decide that its authority over CATV needed to extend beyond simply those issues involving microwave service. The Second Report and Order thus expanded its jurisdiction to cover all CATV systems, regardless of whether or not they were served by microwave relays – representing an increase from 400 systems overseen to over 1,200. Another key provision of the 1966 rules was to restrict the growth of CATV in the 100 largest broadcast markets by virtually banning microwave signal importation there. Imported signals were popular in small to medium broadcast markets since they often were those of independent stations known for sports carriage and extensive movie libraries. Hardly anyone would pay for CATV service, however, if it only duplicated local stations that could be received using a home antenna (the only exception being areas such as valleys where there was signal interference).

The FCC's stated reason for limiting CATV growth in larger broadcast markets was to protect the growth of UHF stations there. Few of the nation's broadcast markets are served entirely by powerful VHF stations; there simply are not enough available frequencies. Most markets have stations representing a mix of VHF and the weaker but more plentiful UHF signals. Some markets rely on only UHF-assigned stations for television service. To the extent that CATV represented a real threat here, it was, once again, because of its ability to import distant signals and allow viewers to bypass local stations. In many cases, especially on the East and West coasts, a microwave relay was not even needed to do this. For example, a community antenna in Manchester, New Hampshire can bring in Boston signals and a community antenna in San Diego can bring in Los Angeles signals.

At this point the FCC was still fairly uncertain about the nature of its authority over CATV – and with good reason. For one thing, its measures did relatively little to help the small-market broadcasters who had initiated the regulatory process by complaining about CATV back in the 1950s. Many of them had gone out of business simply for lack of advertisers in the small areas they served. Among those that had survived that problem, quite a few probably had CATV (or boosters and translators) to thank for extending their audiences. In actual fact, the 1966 rules wound up protecting many corporate-owned major-market VHF broadcasters – who would have done quite well regardless of the regulatory intervention. Most of these stations were also affiliated with (in several cases, owned and operated by) the "big three" broadcast networks. The FCC found itself having to bid farewell to its cherished doctrine of localism in broadcasting, since the last thing powerful network affiliates wished to do was compromise the reliable revenue stream brought by national network programming by replacing it with lower-quality local fare. At this stage, the most promising source of local programming was through CATV, but this would not be given major consideration for a few more years.

In the meantime, the CATV industry was struggling to make the most of a few concessions allowed by the 1966 Report and Order. For one thing, the newer rules reduced the nonduplication period from 15 days before and after a broadcast to just one 24-hour period. For another, the FCC was fairly lenient when it came to local CATV

Box 3.3   The Seiden Report

As of 1964 the FCC was still waiting for what it hoped would be an official Congressional go-ahead to regulate cable. With none forthcoming, however, the Commission hired Dr. Martin H. Seiden, an independent economic consultant, "to compile and analyze the significant factual data relating to CATV and to make recommendations which his research showed to be necessary" (iii). His primary goal was to determine whether CATV did in fact limit the potential growth of UHF broadcast stations and thereby prohibit the formation of a truly national system of broadcast television service. When submitted, Seiden's findings showed CATV to be less of a threat to the broadcast television industry than previously believed.

Seiden looked first at the FCC's stated regulatory priorities (particularly those articulated in the 1952 Sixth Report and Order) as well as the existing economic structure of the US television industry. Clearly he wished to draw comparisons between the Commission's idealistic goal of nationwide broadcast television coverage and the realities of a commercially supported industry. He analyzed proposed CATV regulations, expressing approval that the operations of existing systems generally would be left alone, for this would ensure that rural areas and smaller communities could continue to enjoy quality television service. But after thorough analysis of the likely impacts of CATV on UHF growth in broadcast markets, he judged that CATV did not present as much of a threat as had been claimed. Instead of harsh restrictions on CATV in those markets, he advocated more support for UHF start-ups there. The CATV "problem," he felt, would take care of itself.

Seiden argued in his report that three-network service could only be possible in most medium-sized television markets with some sort of structural assistance that was not yet in place. Until such assistance was implemented, CATV would continue to play a role in providing full broadcast coverage. He argued also that CATV would continue to be the only economically feasible means of delivering three-network television service to markets ranked below number 177 in size – not to mention rural communities. Clearly Seiden saw a need for CATV service. He observed, moreover, that

CATV might actually give fledgling UHF stations a short–term advantage by making their signals available to households without UHF-capable receivers (since that technology was still relatively new at the time). As he explained, "UHF cannot be encouraged by thwarting competition whenever it threatens to appear. It must have a basic viability that precludes the necessity of establishing national policies designed to collect TV homes for UHF audiences" (86). Seiden also found, through extensive interviewing, that while both national and local advertisers were aware of CATV, there did not yet appear to be any reliable means of assessing its impact on their selection of stations or networks as advertising outlets. So although he did acknowledge that CATV had *some* impact on broadcast markets, he didn't feel that eliminating it would actually help the FCC achieve its 1952 goals.

Seiden clearly advocated a policy-making approach that would buttress efforts in UHF television rather than imposing a heavy regulatory burden on CATV, an industry whose economic impact was not yet well understood. This advice was not reflected either in the 1965 First Report and Order or the 1966 Second Report and Order – bodied of regulations that proved harmful to CATV without doing much, if anything, to help broadcast television. The FCC ultimately would expend as much effort amending or overturning a number of those regulations as they had implementing them in the first place.

What if the FCC had adhered more closely to Seiden's recommendations when implementing CATV policy? We can never know, but it is interesting to speculate. On the one hand, giving freer rein to the CATV industry at that stage might have allowed an earlier merging with pay-television and given modern cable an identity truly distinct from broadcast television. Some might speculate that this could have brought an end to "free" (i.e., commercially supported) broadcast television. After all, this was one of the concerns expressed by opponents to early pay-television experiments. On the other hand, strengthening the UHF stations at this early stage might have weakened CATV's position to the extent that modern cable would have developed as nothing more than a utility service (like gas and electric or telephone), existing for no reason other than to strengthen and relay the signals of an array of successful broadcast stations.

systems working out "private arrangements" with local broadcasters. In some cases, CATV systems and broadcast stations were jointly owned, while in others the local stations had relatively few objections to the presence of CATV. In fact, a few of the top 100 broadcast markets were already seeking bids for CATV franchises at this point – not surprising given that CATV's channel capacity in the 1960s far exceeded the number of local stations available in most markets.

Within a short period following implementation of the 1966 rules, the FCC was already realizing those rules were flawed. The Commission actually did not want to bring about the demise of CATV at this stage, since it had come to see the wired medium as the only realistic means of achieving national broadcast television coverage. For its part, though, the CATV industry was in a bind. The inroads it was making into larger communities – those also served by broadcast television – were not enough to garner the sorts of revenues it would need to truly expand the services it could offer subscribers (such as local programming, pay services, and extra channels). Trials of these sorts of services in small communities outside broadcast markets were sufficient to demonstrate their potential, but hardly enough to be profitable. For example, Meadville Master Antenna of Meadville, Pennsylvania began offering a channel of local programming during the mid-1960s. But while that channel made for good public relations, it always had to be subsidized by other operations of the system's owners. The Meadville channel will be discussed further in Chapter 4.

FCC had finally asserted its authority over CATV by the mid-1960s, and the federal agency would have an increasingly important role as the CATV industry grew into the colossal modern cable television industry in the coming decades. But at this point, FCC policy would need to be adjusted considerably before CATV could move forward. The Commission, seeking assurance of its authority in matters relating to CATV, had included in the Second Report and Order a request for Congressional action. While a bill was put forward shortly thereafter, no legislation was forthcoming. In fact, there would be no Congressional action on CATV or cable television until 1984.

Interestingly, a Supreme Court decision in 1968, *United States v. Southwestern Cable*, would serve as the Commission's only formal authorization to regulate CATV until that time. On first manifestation,

Southwestern was not about the matter of the FCC's authority to regulate cable. Rather, in 1960, Midwest Television, a San Diego television station owner, had filed a complaint with the FCC alleging that its (and other stations') fortunes were being jeopardized by local cable system Southwestern Cable's importation of Los Angeles stations. The complaint was similar to those of the western stations discussed above, with the exception that the San Diego case did not involve microwave relays. In 1966, the FCC finally responded by prohibiting Southwestern from expanding its service area. This was a temporary restriction pending further hearings; however the Court of Appeals, at the request of Southwestern, overturned the order, maintaining that the FCC lacked such authority over cable television. The case at this point shifted into a question of the FCC's regulatory authority and went to the Supreme Court. Two years later, the high court overturned the lower court's decision, noting the necessity of regulating the "explosive" new industry, which it did find to conform to the FCC jurisdictions outlined in the 1934 Communications Act. Also of concern was the cable industry's reliance on distant signals, more and more of which were being retransmitted across state lines.

Copyright

Another critical policy issue affecting the future of CATV was settled by the Supreme Court – at least temporarily – on the exact same day in 1968 that the *Southwestern* case was finally resolved: that of programming copyright. It should not be surprising that copyright issues had come to haunt the CATV industry; after all, it relied almost exclusively on television programming paid for by other parties. Broadcast television stations and networks were indeed troubled by this, but unfortunately for them, the most current federal legislation governing copyright at the time was the 1909 Copyright Act. Obviously there were no provisions in this legislation for broadcast programming, much less the retransmission of it by a third party. Copyright actually had not presented many problems for the CATV issue during the 1950s – although attorney E. Stratford Smith recalls how he repeatedly cautioned operators against operations (other than retransmission) that involved copyrighted material (specifically, showing movies or videotaped television programs on vacant channels –

something a number of operators wished to do). As he explained in an oral history interview:

> It was . . . my non-legal and exasperating responsibility to get the industry to hold the line on the master antenna concept. As the years were going by the industry was changing and many were fretting under the yoke of master antenna and to some of them I was a negative force in the industry. But we could not afford to lose that case because every operator would be liable for treble damages for every infringement going back several years . . . Theoretically, hundreds of millions of dollars were at stake industry-wide in treble damage actions and a loss could have, and probably would have, turned control of CATV over to the entertainment industry, and CATV systems could have become the electronic equivalent of motion picture theaters but with no way to sell popcorn. At least that's the way we saw it. In theory, at least, copyright owners could have shut down the operation of a system or even the industry. (Smith oral history)

Smith was not able to fend off litigation forever, though, especially since a number of operators – eager to offer new services to their subscribers – had ceased to heed his warnings. Additionally, the increasing number of CATV systems importing distant signals had begun to raise the hackles of the stations that had originally paid for the imported programs. In 1962 the case of *Fortnightly Corp. v. United Artists Television* (1968) went before the US District Court in New York – where it would remain for the next six years. The Fortnightly Corporation owned two small CATV systems in West Virginia. In the case, United Artists, a major television syndication company, challenged Fortnightly's right to use and profit from its copyrighted programming. The case was a difficult one, monitored anxiously throughout its duration by the entire CATV industry. After all, a decision favoring United Artists could well have put an end to CATV. Initially, United Artists did win the case, and this was affirmed in the US Appeals Court. In the Supreme Court, however, the decision was reversed in favor of Fortnightly Corp. on the grounds that the relaying of electronic signals does not constitute actual "performance" of copyrighted material – as would be the case with a television station or a movie theater.

While CATV's case was helped by the absence of copyright law specific to electronic mass media, it was bolstered even more by the

Figure 3.2   E. Stratford Smith [The Cable Center]

choice of Fortnightly as a defendant. Fortnightly's systems were small and neither used non broadcast-derived programming or imported signals via microwave. Had the related case of *Teleprompter Corp. v. CBS, Inc.* (1974), initiated in 1964, been decided before *Fortnightly*, the verdict might have been entirely different, since, as discussed above, Teleprompter (note the changed spelling) was actively pursuing major-market system ownership as well as program origination. By the time *Teleprompter* was finally decided, the industry was a very different one, and the decision in favor of the defendant probably was a way of delaying policy-making on the cable copyright issue until final passage of a much-revised copyright bill moving through Congress by that point.

In 1968, with the copyright issue settled for the time being, a lot began to change for the CATV industry. Of course not every change can be attributed to copyright specifically, but by the late 1960s, CATV operators felt much more confident in pursuing new programming options, including pay-television. During the years when CATV was in limbo because of FCC regulations and the copyright issue, there had been some developments in pay-television – its technologies as well as its own regulatory scenario. The ability of both CATV and pay-television to expand and develop would have a huge impact on CATV's leap into the modern cable era. This was the point when the CATV industry became known officially as the *cable television* industry. It would not be long before *pay-cable* became an inextricable part of that industry.

Pay-Television

The pay-television industry, which had experienced technical, regulatory, and economic fits and starts throughout the 1950s, made another series of appearances during the early 1960s. Three major system tests were launched shortly after the FCC gave temporary approval to pay-television in March 1959: International Telemeter, Phonevision, and STV. For the most part, these were reconfigurations of earlier systems, but in the 1960s they received a great deal more public attention than they had previously. Of course, the CATV industry continued to pay close attention from the sidelines.

During the late 1950s, International Telemeter had been actively promoting a wired pay system with a coin-box pay-per-view device that presumably addressed subscribers' concerns about having to pay for unwanted programming (which had been a criticism of the flat-rate Bartlesville system). In 1960, the company put this system into operation on an experimental basis in the Toronto suburb of Etobicoke, Ontario. The selection of Etobicoke was very deliberate: Telemeter needed access to theatrical movies (just as Telemovies had). By this point, though, Telemeter was a wholly owned subsidiary of Paramount Pictures, a major Hollywood movie studio. Paramount had been banned from ownership of any US movie theaters following the Justice Department's 1948 Paramount Decree. However, Paramount did hold a major interest in the Canadian exhibition chain Famous Players Theatres. Furthermore, Canadian audiences, long exposed to US media, could be assumed to give a good indication of how American audiences might respond to the pay-television system.

For five years, International Telemeter provided Etobicoke subscribers with an array of movies, sports, series, and specials. And for a while, viewership for these programs surpassed viewership of US network programs. Still, the system eventually began to lose revenue and had to shut down in 1965. Analysts have not been fully able to understand why this happened – though it might have been related to the FCC's awkward and uncertain position on CATV at the time, not to mention the pending copyright decision in the federal courts. Affiliation with existing CATV systems probably was Telemeter's best bet for getting off the ground in the United States, and the company was actively pursuing these sorts of ventures. Unfortunately, most CATV operators were extremely skittish at this stage about becoming involved with anything other than community antenna service – due in part to limited financial resources but even more in fear of jeopardizing the already uncertain legal status of their industry. Since Telemeter's offerings resembled those that Home Box Office eventually would offer, one must wonder how successful this venture would have been without so many obstacles specific to the time period.

Phonevision also made a return by launching a broadcast-based pay system in Hartford, Connecticut in 1962. It offered a program mix similar to that of Telemeter and continued service through 1969 (ironically, the year in which the FCC finally legalized broadcast pay-television on a permanent basis). The Phonevision trial was of relatively little interest to the CATV industry, since there was little chance of it becoming compatible with wired delivery systems. Nevertheless, CATV operators were willing to follow any pay-television experiment to some degree – if for no other reason than to see how alternative delivery systems for televisions would fare.

Of much greater interest was Subscription Television Inc. (STV), a wired pay system tested in Los Angeles and San Francisco for four months during 1964. While the shortest lived of all the 1960s pay-television experiments, STV was also the most prominent. STV represented a substantial reconfiguration of the Skiatron system from the 1950s (changed, not least, by the switch from broadcast to wire). In the intervening decade, several parties from the entertainment industries had become involved with the system, including publishers, electronics firms, baseball clubs, and most notably, former NBC executive

Sylvester L. ("Pat") Weaver.[3] STV's schedule represented a programming mix similar to the systems described above; however, its most prominent attraction was Major League baseball. STV had gained exclusive rights to games of the recently relocated Los Angeles Dodgers and San Francisco Giants (indeed, it is likely that the company's principals had played a role in the teams' relocation from New York to California) (see Whiteside).

Unfortunately for the aspiring pay-television service, even this sort of coup was not enough to keep it in business. In spite of being well backed at the start, STV ultimately fell victim to a well-orchestrated, not to mention underestimated, body of opposition to pay-television. Even before STV had its launch, the Citizens' Committee for Free TV and the California Committee for Free TV had been actively promoting the passage of Proposition 15, proposed legislation to ban pay-television in California. By the time Proposition 15 was passed, in November 1964, STV was already drained of resources from trying to fight it. Even though Proposition 15 eventually was declared unconstitutional, STV was never able to restart its business. Here again was a promising effort to start pay-television in the United States – quite similar to the pay-cable services that would begin within a decade – that wound up being a victim of circumstance.

Why the organized resistance to pay-television? While the opposition groups' names implied that these were grassroots initiatives, mostly this was not the case. Advertiser-supported television, of course, is hardly free, so it was mostly parties with an interest in the continuation of broadcast television and its existing funding structure that led the charge against pay-television (though other groups, such as women's organizations, were also enlisted). These included television networks and broadcast station owners as well as antenna manufacturers (program producers no doubt saw less of a threat in pay-television). It also included movie theater owners who, unlike movie producers, would see no profit from the showing of movies on television instead of in theaters.

The concept of wired pay-television by no means disappeared with STV, however. In fact, the mid-1960s were the stage at which the pay-television industry and the CATV industry were beginning in earnest to explore joint ventures. While it would be a few more years before these projects reached fruition, there are several examples of what was in the works. International Telemeter, after closing its

Etobicoke operation and abandoning plans for similar systems, actively lobbied the FCC to authorize pay-television as a part of CATV service. Unsuccessful in this effort, the company eventually moved into the manufacture of multichannel components for CATV systems – obviously hoping that the "supply" of more channels would lead to consumer demand for programming to fill those channels. After yet another corporate reconfiguration, STV also pursued the development of pay-television systems that would be compatible with CATV. Even though neither company actually accomplished these goals, they can at least be credited for pursuing the notion of pay-cable during that medium's formative years.

Other nascent ventures, including the experiments by Teleprompter discussed above, would prove more successful in paving the way for pay-television as we know it today. Although the notion of broadcast pay-television did not disappear entirely with the rise of pay-cable (and would even resurface briefly during the 1980s in the form of a handful of short-lived businesses in urban areas), it was clear that wire-based forms of pay-television were going to dominate the market. Within the next few years, almost the entire concept of pay-television would become synonymous with pay-cable.

Another emerging technology that would have a huge impact on CATV and its convergence with pay-television was communications satellites. When joined with pay-television, satellites would revolutionize the industry in such a way that the term "CATV" would become obsolete almost overnight. The new emphasis would be on new technology and its potential, making terms like "cable television," "satellite cable," and "wired nation" the buzzwords of the day. As will be discussed in detail in Chapter 4, satellites stood poised and ready – at least in the public's mind – to transform electronic communication from the ground up.

Satellites

Sputnik, the first artificial satellite, was launched by the Soviet Union in 1957. This was at the height of the Cold War and, even though *Sputnik* really did nothing more than transmit periodic beeps back to earth, it represented a significant advance in telecommunications technology. It also heralded the start of the "space race" between the Soviet Union and the United States – a period in which the two competing superpowers enlisted science and technology to demonstrate economic, military, and political prowess. More than anything else, however, the competition probably did more to advance domestic telecommunications in each nation (and before long, in other nations as well) than to actually prove either nation more technologically advanced than the other. The United States responded to *Sputnik* by launching the Army's *Explorer-I* satellite a year later. Used primarily for gathering atmospheric data, *Explorer-I* is best known for detecting the Van Allen radiation belt that encircles the Earth.

The US launch of Bell Labs' *Telstar* satellite in 1962 was more of a breakthrough for the telecommunications industries specifically. With the first *active* transponder, *Telstar* was able to transmit the first intercontinental television broadcast, a press conference by President John F. Kennedy that was viewed in France. *Syncom 2* came a year later. It was the first geosynchronous and geostationary satellite, meaning that its orbit matched that of the earth's rotation, making it appear stationary and demonstrating that eventually satellites would be able to transmit regular program schedules. This capability would be the breakthrough that, a little more than a decade later, would allow cable-specific channels (networks) to present viable competition for broadcast television.

In the intervening period, government policy-makers began to ponder the implications of the new technology. In 1962 Congress passed the Communications Satellite Act, establishing the Communications Satellite Corporation (COMSAT). COMSAT was a public corporation intended to develop a commercially based international satellite communications system; it was owned by some of the major corporate players in satellite communications in a joint venture with some independent investors. Somewhat reminiscent of how the government had sanctioned the creation of RCA in 1919 to both nationalize and expedite the new technology of radio broadcasting, it created COMSAT in order to develop satellite technology as quickly and competitively as possible (though it should be noted that the government's role in the latter was much more direct and deliberate).

The process did not move all that quickly – certainly not as quickly as the television industries would have liked. Broadcast networks were the first to approach the FCC about satellites, since they saw the new

Figure 3.3   *Syncom 2*, launched in 1963 [NASA]

technology as a possible means of bypassing the expensive AT&T land lines when transmitting programs to their affiliate stations. But rather than give immediate approval to an application by ABC and Hughes Aircraft Corporation in 1964, the Commission (and other federal government bodies) initiated a series of meetings to determine the best ways to finance and regulate a commercial satellite system (again, similar to the series of radio conferences held in the 1920s, following the creation of RCA and prior to the passage of the 1927 Radio Act). The Johnson White House took a strong interest in the issue as well, and a great deal of satellite policy research was carried out under the auspices of a presidential Task Force on Telecommunications Policy led by Eugene V. Rostow, former Undersecretary of State for Political Affairs. Other policy studies also fed the discussion, leading (as discussed in Chapter 4) to a great deal of optimism surrounding satellite technology. Nonetheless, a policy on domestic communications satellites would not actually be put in place until 1972, under the Nixon administration.

The International Scene

Other nations also were contemplating the implications of satellites and other new technologies for their own present and future electronic media industries. Surely satellites would be able to replace some existing delivery technologies there as well. For example, the relay services discussed in Chapter 2 continued to flourish in the UK and some other western European nations, though they did not develop into CATV as they had in Canada. Satellites held the potential to extend the reach of national broadcasters even more efficiently. By the time satellites became available for this role, however, they would already be the distribution medium of choice for the array of private television providers that would challenge the very existence of those national broadcasters.

Meanwhile, by the end of the 1960s, Canada had its own highly developed CATV industry – that would grow in tandem with that of the United States. Canadians enjoyed watching American television, and since over half the communities in that nation are located close enough to the US border for easy community antenna reception of those signals, there was much incentive for amateur experimentation with CATV there. In fact, so great was the demand for US programs via community antenna that CATV, once off the ground in Canada, had grown faster and gained higher rates of penetration than it did in the United States (or anywhere else in the world). Of course the rapid development of CATV in Canada was helped at this stage by the fact that no regulatory body had imposed restrictions of the sort that existed in the United States.

Canada's CATV operators had also developed their own trade association, NCATA (National Community Antenna Television Association of Canada), in 1957. Like the NCTA, NCATA was formed both because of regulatory concerns and to share technology and business strategies. As in the United States, the Canadian CATV industry grew more concentrated in the 1960s, especially as large cities (besides Montréal, which had been wired for CATV a decade earlier) presented themselves as fertile ground for new systems. Famous Players, the

Canadian company behind International Telemeter's Etobicoke pay-television trial, became an active developer of CATV once it saw how popular and profitable the technology could be.

It would not be long though before Canada's growing cable television industry would face its own set of regulatory challenges – precisely *because of* the heavy demand it was fostering for American television programs. While CATV in the United States performed the rather necessary function of making television service available where it would not have been otherwise, the situation was different in Canada. The CBC, in the tradition of the BBC, strived to provide near-universal coverage of broadcast television – even in very remote and isolated sub-arctic regions. In other words, few Canadians would have lacked television service without CATV. Rather, CATV made it easy for Canadians to bypass programming produced within and for their own nation in order to watch the heavily commercialized American fare that was pouring in. Canadian policy-makers, not surprisingly, grew very concerned about the loss of national identity this seemed to be creating.

Canada's situation is telling. In virtually every national context, the arrival of multichannel television (at whatever point this occurs) dramatically alters the television institutions already in place – most notably by introducing large amounts of advertiser-supported and foreign programming, and thus creating new standards and expectations for what is produced and aired locally. This has been a mixed blessing, to say the least, since outside the United States, imported programming most often refers to programming from the United States. While this programming is entertaining and boasts very high production values, it does not, by and large, reflect the values held by other national governments regarding the media. One cannot overlook the fact that the United States is one of relatively few countries in the world in which the broadcasting industries (both radio and television) developed almost exclusively through commercial revenues, with no support from public funds. Most other nations chose either direct or indirect government oversight of their broadcast media industries in order to bolster indigenous media production and strengthen national, regional, and ethnic forms of expression.

By the 1960s, the problem of cultural imperialism was becoming quite perceptible all over the world, in fact – a problem that would only grow as new technologies made it cheaper and easier to distribute media content across national borders. Although Canada has struggled with this issue, it has been better equipped to combat it than some other countries; Canada at least has had the resources to develop and finance its own media infrastructure. Many developing nations, whose economies have been unable to support much (if any) indigenous television production, have resorted to importing ready-made programming from the United States and other Western countries. This happens because of a very basic economic fact: television programs (like all media products or "software") are produced once and reused indefinitely, so once a nation's television production operations have become self-sustaining domestically, exporting programs will be extremely profitable regardless of how much is charged to those importing them. This deceptive "win–win" arrangement has caused many problems, however. The lifestyles portrayed on Western television programs are not even reflections of life in the countries where those programs originate, so the values they espouse can be entirely alien (not to mention unrealistic and often degrading) to other countries.

Has US *cable* specifically had a role in this sort of cultural imperialism? It surely has. The more channels that are available domestically, the more incentive producers have to turn out new programming. Then the greater the supply of programming available, the more there is for export. Moreover, during the last decades of the twentieth century, the technologies designed to facilitate television distribution in the developed world would allow that same programming to flow into less developed countries – making it easy for those countries to bypass the important stage of forming their own, indigenous television production infrastructures. In the remaining chapters of this book, it should become apparent that each stage of technological development in multichannel television has brought with it a range of complex policy issues, both in the United States and internationally.

NOTES

1 Unlike other urban areas, the need for CATV had arisen in Manhattan because the numerous and closely clustered skyscrapers impeded broadcast signals in ways similar to mountains.
2 An interesting consequence of this was to allow many of the programming innovations that would define modern cable television to occur in small

CATV communities rather than major metropolitan areas. Manhattan was a notable exception since its CATV systems predated the FCC regulations.

3   Weaver, who had served as programming president and later chairman of NBC, was known for developing the *Today* and *Tonight* shows as well as the spot or "magazine" format of program sponsorship still in use today. His role in STV no doubt lent a degree of credibility to the system and to pay-television in general.

FURTHER READING

Mullen, Megan. "The Pre-History of Pay Cable Television: An Overview and Analysis." *Historical Journal of Film, Radio, and Television* 19:1 (March 1999), 39–56.

Seiden, Martin H. *Cable Television U.S.A.: An Analysis of Government Policy.* New York: Praeger, 1972.

Whiteside, Thomas. "Onward and Upward with the Arts: Cable I-III." *New Yorker*, May 20, 1985, 45–85; May 27, 1985, 43–73; June 3, 1985, 82–105.

Chapter 4

"Blue Sky": 1968–74

Virtually nothing in the US regulatory model has ever provided for equal availability of television service throughout the nation and across different cultural groups. An advertiser-supported system is almost never able to serve areas of low population density. Catering to the expectations of advertisers can never assure the diversity of programming needed to address the needs and interests of the entire population. And an advertising-dependent system can never offer an adequate balance of educational/informational and entertainment programming. In the twenty-first century, people have grown rather complacent on this issue; some are even pleased with the situation, seeing it as evidence of free enterprise at work. But in the 1960s there was more zeal about overhauling US television.

Beyond the world of television itself, this period in US history saw a great deal of social upheaval, with the civil rights and women's rights movements under way as well as the Vietnam conflict and protests against it prominent in news headlines. It was also a period of technological optimism and competitiveness – especially as manifested in the "space race" against the Soviet Union, which continued to feed advances in satellite technology. In television programming history, the period saw increased amounts of broadcast news, the founding of the Corporation for Public Broadcasting (CPB) and the Public Broadcasting Service (PBS), and the rise of "socially conscious" situation comedies such as *M*A*S*H* and *All In the Family*.

In cable history specifically, this period was known as "Blue Sky" because of the expectations suddenly placed on that medium to become a force for social change. At this stage of television history, with

information circulating about the capacities of cable-related technologies such as pay-television and satellites, cable began to be perceived as a beneficial supplement – or even alternative – to broadcast
television. And its numbers continued to grow, with 2,490 local
systems and 4.5 million subscribers as of 1970 (*Television & Cable
Factbook*). If the first two-thirds of the 1960s witnessed the imposi-
tion of federal regulations that brought the CATV industry's growth
to a near standstill, the last third of that decade saw quite the oppo-
site trend. From 1968 to 1970, government policy-makers – along
with the general public – appeared to be doing everything possible to
ensure that cable television (as it was called by this stage) would
continue to flourish. Rather than being perceived as the destroyer of
broadcast television, cable had come to be seen as the remedy to all
of that predecessor medium's ills. Obviously this was a sudden change
of sentiment. So what happened? As this chapter will explain, the
various social and political developments of the time introduced new
ways of thinking about cable television's potential in the United States.
New technologies were being developed and cable reform proposals
circulated. And new policies were being implemented rapidly in response to all of this.

"Blue Sky"

Blue Sky is a broadly encompassing term referring to numerous commissioned studies, newspaper and magazine articles, and popular
sentiment aimed at taking full advantage of a satellite-served cable
television industry. It also refers to the policy measures taken in response to this discourse. Finally, it refers to the shift in perceptions
about cable television brought on by the combination of all of these
documents and other expressions of optimism. In the minds of policy-
makers, cable went from a regulatory conundrum, even a burden, to
a source of salvation from poor broadcast policy planning. For the
public, cable went from being completely unknown to all but the tiny
portion of the population using it for basic television service to promising a "cornucopia" of additional channels available to everyone.

The FCC regulations passed in 1965 and 1966, in spite of their
restrictive provisions, had called some attention to the existence of
CATV, and because of this some articles appeared in the popular

press attempting to explain how the medium functioned and why it had been brought under regulatory scrutiny. These were fairly basic in their approach to the topic. However, it was not long before more visionary pronouncements began to appear as well. These were fueled in part by the fact that some cable systems operating in the New York City borough of Manhattan – where the tall buildings blocked broadcast signals in much the same ways that mountains did – offered a sort of laboratory for what cable could offer in areas with potentially high subscriber numbers. The Manhattan systems, whose launch had predated the FCC's mid-1960s rule-making by just months, had been allowed to continue operating under a grandfather clause once cable had been banned from virtually all other top 100 broadcast markets. A task forced organized by the New York mayor's office and headed by former CBS News president Fred Friendly recommended at the end of 1968 that the city's three operating cable systems begin producing and airing their own programming – which they soon did, offering such locally oriented programs as *The Community Bulletin Board*, *Manhattan Issues*, and *Town and Village News*, along with various movies from Hollywood's vaults.

The report issuing from Manhattan only echoed other, more broadly applicable pronouncements, though. Actually a wide spectrum of constituencies across the United States had begun to embrace cable's potential for television reform by this point. At the top level, President Johnson launched the Task Force on Telecommunications Policy under the direction of Eugene V. Rostow, as mentioned in the previous chapter. Even though the Task Force considered a variety of means by which to unlock television's three-network oligopoly – including an enlarged array of UHF stations, a fourth network, and direct broadcast satellite – cable emerged as a clear favorite. The group's final report ("The Rostow Report") cited cable's high bandwidth, ability to charge viewers directly for service, and established industry structure as deciding factors. The Rostow Report also advocated the creation of an entirely new federal agency charged with coordinating current and future telecommunications technologies.

Other commissioned studies on cable's future were more focused on how the industry might be structured by existing government bodies so as to derive the greatest benefit from market-driven forces. For example, the Sloan Commission (see Box 4.1) advocated revised copyright legislation as a replacement for the restrictions on distant

Box 4.1 The Sloan Commission Report

The optimism about cable's potential, writ large, subsumed any differing ideas as to what its specific content or purpose should be – and this is likely to account for its strong influence on policy-making at this point in time. As sociologist Thomas Streeter put it: "Diverse and often antagonistic viewpoints were united around a shared sense of awe and excitement; maybe the new technologies were good, maybe they were bad, but in any case they inspired a sense of optimism and opportunity" (222). However, as Streeter also makes clear, each of the parties engaged in Blue Sky had its own particular agenda and set of goals for a restructured and re-formed model of television service. The Alfred P. Sloan Foundation is one example, with its report, formally titled *On the Cable: The Television of Abundance*, being one of the best-known and most widely circulated Blue Sky documents – at one time available in nearly every library in the United States.

The Sloan Foundation is the legacy of the founder of General Motors. Its longstanding philanthropic emphasis has been the social impacts of technological change, so its engagement with the Blue Sky discourses should not be surprising. In June 1970, the Foundation's trustees established the Sloan Commission on Cable Communications, charging its members with making recommendations for the future development and uses of cable technology in the United States. The following segment from the Commission's final report epitomizes the sentiments of the Blue Sky era generally, for it, like so many other writings of the period, situated cable as the potential remedy for a system of broadcast television that seemed to have failed the American public:

> If one has any faith at all in the value of communications, the promise of cable television is awesome. The power of the existing system [of broadcast television] is immense; it dwarfs anything that has preceded it. Never in history have so many people spent so much time linked to an organized system of communications. But where it has dominated communications in power, it has become almost trivial in scope. It has dealt primarily with entertainment at a low level of sophistication and quality, and with news and public affairs at their broadest and their most general. It has been obliged to think of the

mass audience almost to the exclusion of any other, and in doing so has robbed what it provides of any of the highly desirable elements of particularity. (167)

This was typical of Blue Sky rhetoric, but a closer look at the Sloan Commission report also reveals a fairly conservative, private enterprise-minded approach to cable's development. This actually was very much in keeping with the philosophies of Alfred P. Sloan himself, who is known to have orchestrated effective anti-public transit campaigns in the United States during the early part of the twentieth century (targeting electric streetcars particularly) as a way to promote his company's gasoline-powered vehicles.

Viewed another way, however, some aspects of the Sloan Commission make its approach seem far more realistic than some other Blue Sky writings in its approach to cable's development – especially in retrospect. They recognized the important role of a technology's end-users, stating: "The shape [cable's future uses] will assume will be determined on the one hand by entrepreneurs, public and private, who are willing to take the responsibility for risking money and career on the promotion of an idea or an ideal, and on the other hand by the users of the system through the response they make to the undertaking of the entrepreneur" (10).

The Commission, like other Blue Sky visionaries, had its ideal scenario – speculating, for example, that:

Like the press, it can be directed toward a wide variety of uses. The press deals largely with entertainment, in the form of books, of magazines, and to a large degree of newspapers. In these same forms, it is the basic medium of information. It is a vehicle by means of which education, formal and informal, is conducted. It carries the burden of offering an outlet for opinion, warranted or unwarranted, popular or unpopular. It is an essential intermediary in the provision of almost every kind of service, private and public. It is irreplaceable in the political process. (43–4)

Later, the Commission said that a "cultural channel" for arts aficionados:

[m]ore or less identifies itself. The problem for the cable television entrepreneur will be to define other audience groups, comparable in

size to or larger that the cultural audience, to whom special appeal may be made and for whom special programming may be designed. The audience for news, documentary programs, public affairs and the expression of opinion is one such group that comes immediately to mind, but perhaps falls outside the area of entertainment and involves other problems ... There is also the audience that is represented in the world of magazines by *The New Yorker* and the like. There are the special audiences of young mothers, of the aged, of teenagers. One might consider a "professional channel," appealing in alternating time segments to doctors, lawyers, educators, engineers. One might well consider an "ethnic channel" appealing in alternating time segments to the various ethnic sub-groups in the community. Many of these channels would intermix entertainment, news and public affairs programming. (69–70)

But the Sloan Commission also asserted that it was unrealistic to expect any Blue Sky scenarios to materialize overnight. Its report discussed at length a transition period, during which cable would need to rely on readily available and inexpensive programs such as off-broadcast reruns. It felt that effective copyright legislation was in order as a means of ensuring fair compensation for programs used in such a scenario, and that existing restrictions aimed at protecting broadcasting interests should be lifted right away. To force more rapid development of alternative programming models could only lead to long-term failure – both economically and in terms of shaping public tastes.

The Sloan Commission felt that: "In the end, cable must grow as conventional television has grown: on the basis of its own accomplishments. As it takes on an identity of its own, the current debate over distant signals and the passion it arouses, as well as the disputes concerning the rights over local broadcast signals, will come to appear insignificant stages in the growth of a total television system" (62). The Sloan Commission's plan for cable was closer than those of many other Blue Sky parties to the free-market stance ultimately taken by US government regulators with regard to cable television – even if it was not a complete match. And today's multichannel television environment might well be described as the sort of programming mix the Commission had envisioned. There are still many reruns, but more and more cable networks are producing or acquiring their own original programming – often at a cost competitive with the long-established broadcast networks. Analysts must ponder: Did the Sloan Commission merely predict what would have emerged in a free-market scenario regardless of any specific recommendations? Would the already deeply entrenched viewing habits and preferences of the US television audience have even permitted a more radical shift to take place?

signal importation. They also maintained that a prohibition on multiple-system ownership would help to foster local programming operations. The RAND Corporation, in a series of studies funded by the Ford Foundation, made essentially the same argument for copyright reform and the lifting of existing restrictions on the cable industry.

Articles also appeared in the popular press. For example, FCC commissioner Nicholas Johnson published an article in *Saturday Review*, in which he pointed out that:

A cable system could, if so designed, reach precisely selected geographic portions of a city – or the nation – which may correspond to particular social, economic, or other special interest groupings. Cable could become a viable medium for interconnection of what would, in effect, be a number of large closed-circuit systems. Whereas a local broadcaster may not be able to justify programming aimed just at ballet enthusiasts, or the local Negro community, or *aficionados* of sports cars, a regional or even a national cable network might be developed which could enhance its appeal significantly through specialized programming. (88)

Johnson was not far off in predicting the array of cable channels that would be available by the 1990s – even if his vision was far more public service-oriented than what actually emerged under lightly regulated market competition. He was well aware of the challenges to be faced by anyone trying to dislodge the market-driven television program selections to which the American public had grown accustomed. People tend to be idealistic in expressing what we wish could be available to us, but when we actually choose what to watch on television we tend to resort to the familiar.

Figure 4.1   Sloan Commission Report [author photo]

Journalist Ralph Lee Smith expressed a slightly different vision for cable two years later in an article in *The Nation* entitled "The Wired Nation" (which he subsequently expanded into a book by the same name). Like Johnson, Smith believed cable's potential was being wasted while it did nothing more than retransmit broadcast programming. He imagined cable as a system of common carriers, wherein any entity – ranging from community groups to independent programmers to those wishing to offer data services – could lease access to the wires. Actually, Smith's ideal nationwide cable system, which he presciently referred to as an "electronic highway," bears a striking resemblance to today's Internet. He saw cable as offering a much wider range of communication services than television programs alone, including home libraries, electronic mail, and virtual travel.

Johnson and Smith were among the most prominent Blue Sky writers in the popular press; however there were many other voices as well – touting cable's advantages for every imaginable audience. Educators' organizations proclaimed that cable could benefit classrooms, including the delivery of lessons to bedridden pupils via television. A variety of entertainment trade organizations predicted new job opportunities for their members. And political parties of all sizes and positions imagined virtually unlimited airtime for publicizing their views.

The cable industry itself got some much needed reinforcement during this period, beginning with a speech by Nicholas Johnson at a 1967 NCTA regional meeting in Philadelphia. Of course, Johnson was advocating local programming first and foremost, but one of his recommendations for making that feasible was the lifting of existing restrictions on cable operations. It was Irving Kahn of Teleprompter, the largest MSO at this point, who was the first major cable player to latch onto this sort of pronouncement in plugging his industry while testifying before Congress in 1969. Policy analyst Thomas Streeter has pointed out that, not only did Kahn draw from the optimistic Blue Sky discourse to push for a lifting of restrictions, he went so far as to exploit the climate by declaring that cable would level and democratize the playing field by making spectrum available to a broad range of "impartial" experts (231).

Even though its members clearly were more optimistic about their future prospects than they had been just a few years earlier, the cable industry was still on shaky ground during the Blue Sky period. It is important to remember that all of the proposals for change were being put forward within a vacillating policy climate. The *Fortnightly* copyright case had only just been settled in the courts, even while *Teleprompter v. CBS* was still making its way through. It would not be until passage of the 1976 Copyright Act that the cable industry would have a clearly defined path to follow in the copyright matter. Meanwhile the FCC was already making adjustments to its 1966 Report and Order and trying to resolve some other cable-related issues.

Basically, three sets of FCC rules enacted between 1968 and 1972 would enable the rather rough transformation of CATV into cable television as we know it today: the 1969 Report and Order on Cable Television, the 1972 Report and Order on Cable Television, and the

1968 Report and Order on Pay-Television. The cumulative effect of these rule-making efforts was at first to tighten the FCC's grip on cable television's growth and then subsequently to loosen that grip quite a bit (though not completely). As a federal agency and part of the executive branch of the US government, the FCC's composition and regulatory temperament echoed the shift between the Johnson and Nixon administrations – the latter leaning heavily toward deregulatory policies in various spheres.

Local Origination

The ill-fated 1969 Report and Order on Cable Television was the FCC's first attempt to bring one of the Blue Sky scenarios, local programming, to fruition. It mandated that systems with over 3,500 subscribers begin to originate live local programming prior to April 1, 1971. In other words, medium to large-sized systems would have to come up with some means of *producing* their own programming. For a while there seemed to be some promise in this requirement. Those systems that attempted to comply with the rule offer some fascinating stories of ingenuity, volunteerism, and homegrown talent. As *Broadcasting* reported:

> Many residents in Warner Robins, Ga., a community of 32,000 outside Macon, were able last week to tune in to a new cable-television channel that delivered "grass-roots" live programming.
> Beginning Monday (Nov. 2) at 7:30 a.m., they watched a former speech teacher giving international, national and community news; a local woman exhorting viewers to join in an exercise class; high-school students discussing their activities; a secretary reading stories to children and singing to musical accompaniment; and the mayor of Warner Robins reporting on community issues. (Big-time venture)

A similar story appeared to be unfolding in Honesdale, Pennsylvania, where one of the nation's earliest systems was still being operated by one of its founders. As the trade publication *TV Communications* reported:

> Of course, equipment alone doesn't make good TV programming. It takes people. Strangely enough, [Honesdale TV Service] got a lot of help from an undertaker. One of the first in the community to volunteer to help the new channel was Frank Myers, a local mortician.

Frank has always enjoyed electronics as an avocation. When he heard that HTVS was going to produce local TV shows, he thought it would be a lot of fun to participate. Frank helped [Chief Engineer] Earl Wilcox set up the equipment and he regularly acts as a cameraman.

Talent for in front of the camera was also available on a volunteer basis. Interview shows are often emceed by an art teacher from the local high school. A salesman for a petroleum distributor provides commentary for telecasts of local sporting events like horse races and stock car races.

A very important volunteer was a young lawyer who very capably handled political interviews. He was encouraged by his father, a Wayne County Judge, who said that today, any man with political ambition must know how to be comfortable in front of a TV camera. Therefore cablecasting is the ideal training ground. (Cantor)

Honesdale and Warner Robins – along with several other local cable systems – were resourceful in attempting to comply with the FCC rules. Other success stories tell of how local cable systems went into partnership with public schools or local colleges in order to comply with the rules. Surely there was no shortage of young people wanting to learn television production skills. Based on news coverage of program origination set-ups such as the ones discussed here, one might go so far as to cite them as evidence of the success of the 1969 Report and Order.

However, on a larger scale the 1969 rules were a fiasco. While reversing previous policy that limited cable's ability to do more than provide community antenna service, the rules ended up imposing a strikingly heavy logistical and financial burden on the industry. The FCC simply had asked for too much in too short a time frame. Most programs of the sort described above were considered amateurish by their potential viewers, had low production values, and struggled to find local sponsors. Even with the greatest of ingenuity, they quickly drained their systems' resources. Moreover, the vast majority of cable systems had found themselves unable to comply with the rules at all, and hence had ignored the ruling – frustrating regulators as well as their own ambitious industry.

The FCC had based its requirement on an estimate by equipment manufacturer Telemation Inc. that even elaborate color programming facilities could be feasible for most cable systems. However, in a 1972 study economist Martin Seiden expressed a different – and, as it turned out, more realistic – view, based on such factors as equipment and labor costs for programming, other costs in running a cable system, and the potential for local advertising revenues. Seiden explained:

> If implemented, the FCC's 1969 rule will result in major budget changes for CATV owners. Equipment expenditures and operating costs will represent a new outlay while advertising revenues will represent an unknown income potential. But the implementation of this new requirement, which the CATV industry has long been advocating, is sobering. There are unique possibilities to CATV in its local orientation including local cable networks that if approached with imagination could be made profitable. As indicated [in this study], however, program origination is not going to be economically rewarding for most systems. . . . [I]n all likelihood cablecasting will be a drain on the current profits of all but the largest systems. (31–5)

Seiden was no less prescient here than he had been in his 1964 report on the overstating of CATV's potential harm to the growth of UHF broadcasting (discussed in Box 3.3). Clearly the new regulations needed some review – and this was already in process by the time he published his study.

Cities and Franchising

Meanwhile, the municipalities in which cable systems operated, and their local government bodies, had grown more interested in cable operations. Of course the systems had always relied on access to public rights of way, but in the early years the main obstacle had been access to the utility poles – not challenges from local authorities. But as more and more people were flocking to cable, and as more and more revenue sources lay on the horizon for the cable industry, municipalities were becoming more interested. For one thing, franchise fees presented a lucrative revenue stream for public works projects – and at this stage the only cap to such a fee was what the market would bear in terms of what a cable operator considered reasonable to pass on to subscribers as a portion of their monthly bills. For another, there was virtually no regulation of the cable industry at the state level since it had been determined in the courts that cable did

Box 4.2   George and Yolanda Barco and Meadville
Master Antenna

The father–daughter team of George and Yolanda Barco were not
among the very earliest CATV operators, since they did not launch
their Meadville, Pennsylvania system until 1953. But once they
became involved with CATV they made a huge impact on the
growing industry. Both Barcos were practicing attorneys during the
early 1950s, and it was because of their law practice that they first
learned about CATV. While in New York City on a business trip,
George Barco observed the master antenna system that was sup-
posed to deliver television to his hotel room. Since the television
wasn't working properly when he tried to use it, Barco had the
chance to ask a repairman how the system operated. Upon finding
out, he began to speculate as to how such a system might serve an
entire community. It did not take him long to find out about
CATV systems that already existed at the time – or the newly
formed NCTA.

The Barcos used money from their law practice to launch Meadville
Master Antenna, a CATV system, to serve their northwest Pennsylvania town of approximately 20,000. In the late 1950s they were
joined in this venture by George Barco's other daughter, Helene
Barco Duratz, who served as office manager, and her husband
James Duratz, who served as general manager. In an industry known
for its "mom 'n' pop" businesses, this was truly a family-run operation. The Barcos joined the NCTA and quickly became some of its
most active members, contributing their legal expertise in particular
as the organization grappled with such early problems as the federal
excise tax. In addition to their affiliation with the national organization, the Barcos were founding members of the Pennsylvania Cable
Television Association (now called the Broadband Cable Association of Pennsylvania). They were also instrumental in developing
what is now the Pennsylvania Cable Network (PCN), a cable/
microwave-carried education and public affairs network with programming for and about Pennsylvania specifically.

Meadville Master Antenna offers an important lesson in the economics of cable television since it was run in a way most cable
operators would consider far less profitable than ideal. While the

Barcos invariably used the latest technologies for their system, increasing channel capacity as often as feasible and adding other "bells and whistles" as they became available, they also recognized a need to serve their community. They supported local charitable causes generously and, even more interestingly, subsidized a channel of local programming starting around 1963.

This was several years before the FCC tried to mandate such programming – and the channel, CTV-13, remained in place for decades after the requirement proved unsuccessful. It offered special events coverage, routine coverage of public meetings and high-school sports, and a schedule of studio-produced programs (talk shows, exercise shows, children's shows, and so on) featuring local residents. CTV-13 never made money, but as James Strickler, who managed its operations, said in an interview:

> Initially and throughout the total operation with my being there, the Barco family subsidized origination very heavily. We never reached a point where our advertising covered our expenses. It was an expensive proposition. And Mr. Barco would say to me, "Do the best you can. Don't lose any more money than possible." And we would try to do that, but it was difficult to sell the quantity of advertising to cover all of our expenses . . . Because it wasn't profitable, many of the [other cable] operators ceased [the program origination] operation or didn't try [in the first place]. But Mr. Barco and Miss Barco had a different approach. They felt that it was a "give back to the community" type thing, and they subsidized it in that vein. Most [other] operators were looking at the bottom line figures . . . Mr. Barco often times, in our meetings, would say, "It's lucky I'm still a practicing attorney so that I can afford all this." (James Strickler, interviewed by the author, Meadville, Pennsylvania, July 2004)

Meadville's local cable channel demonstrates what could have come about nationwide during cable's Blue Sky years. With adequate funding, many communities could have had a local channel owned and professionally staffed by their cable operators. However, Meadville also shows exactly how uncommon a phenomenon "adequate funding" actually was. CTV-13 always lost money for its owners – who happened to be able to support it through other income sources, most notably their law practice.

Had the FCC truly wished to have local programming play a role in cable's future, much larger structural changes in the industry would have been necessary. Indeed, the Commission would have needed to address the advertiser-supported nature of US television in its entirety – especially the deeply entrenched broadcast model. What the Barcos accomplished in Meadville was simply unfeasible economically for almost every other independent cable operator. And the wealthier MSOs, increasingly in control of local cable systems, unfortunately lacked the Barcos' dedication and connection to the local community.

not fall under the jurisdiction of public utility commissions. And also, the municipalities were perceiving – much as any other constituency at the time – that cable could have some direct benefit to them. Indeed, one of the more popular functions of local cable programming was, and continues to be, coverage of city councils, school boards, and other local government bodies. Local governments had an interest in seeing this sort of programming continue – financed by cable revenues, not taxpayer dollars. Still, the local franchising process was a fickle one, often subject to the whims of corrupt local politicians. Thus it would be just one of the issues addressed by the next major body of FCC rules affecting cable.

The 1972 Rules

In 1972 the FCC passed yet another Report and Order on Cable Television, this one much more extensive than the preceding ones. The goal this time was to encourage the growth of cable television by relaxing some of the restrictions that had kept the industry from making enough money to implement new services (most notably local programming operations). The Commission clearly had paid attention to what so many of the Blue Sky documents had recommended. Also by this stage the FCC was under the control of the Nixon administration and Chairman Dean Burch – policy-makers far more oriented toward laissez-faire capitalism than those in the Johnson administration had been. The 1972 rules were one of the measures, indeed the most significant of the decade, that they would take in opening the cable business to free-market competition.

The FCC saw one of its first orders of business as asserting its authority to regulate cable television. It required that any system wishing to begin service apply for a "certificate of compliance," and that any existing system acquire such a certificate within five years. Furthermore, the Commission would not issue such a certificate unless the cable system had been franchised in the municipality where it intended to conduct business. The local franchising process itself was formalized by the rules, with the intent of protecting municipalities, cable subscribers, and the cable companies themselves. Franchise fees were capped at 3 percent of revenues for most systems and in no case more than 5 percent. This was significant since, as cable companies increasingly were bidding to wire larger cities, competition was fierce and easily subject to corruption. In fact, Irving Kahn's conviction and subsequent prison term for bribery during the franchising process in Johnstown, Pennsylvania eventually would lead to the demise of Teleprompter Corp.

The rules specified a complex must-carry obligation known as "anti-leapfrogging." At the most basic level, cable systems were obligated to carry signals representing all three broadcast networks. These signals also had to come from the network affiliate stations in closest geographic proximity to the cable system. In other words, cable systems could not bypass network signals originating within their own market areas in order to bring in the signals of larger and (presumably) more popular stations in distant markets. If the signals of independent stations were to be imported, as was desirable due to the additional programming choices they offered, these needed to originate from one of the two largest markets in close proximity to the cable system. Cable systems in the top 50 broadcast markets were allowed to carry signals of three network affiliate signals and three independent stations. In markets 51–100, only two independent station signals were allowed, and only one in markets below 100.

Broadcasters were further protected by program exclusivity provisions in the 1972 rules. A cable system was not allowed to show a program via the transmission of a distant station it carried if a local station held exclusive rights to the program in that particular market. Even though it set a lasting precedent within the cable industry, this rule would prove extremely difficult to follow as well as to enforce.

Table 4.1 Top 100 US broadcast markets, 1972

A broadcast market encompasses not only the city or cities for which it was named, but also any communities lying within the grade A contour of any stations located there. A look at the top 100 markets as of 1972 shows just how much the cable industry's growth was hampered by the 1969 Report and Order and how much it was helped when the 1972 Report and Order permitted cable in markets 51–100.

1	New York, NY–Linden–Paterson, NJ	27	Columbus, OH
2	Los Angeles–San Bernardino–Corona–Fontana, CA	28	Tampa–St. Petersburg, FL
		29	Portland, OR
3	Chicago, IL	30	Nashville, TN
4	Philadelphia PA–Burlington, NJ	31	New Orleans, LA
		32	Denver, CO
5	Detroit, MI	33	Providence, RI–New Bedford, MA
6	Boston–Cambridge–Worcester, MA	34	Albany–Schenectady–Troy, NY
7	San Francisco–Oakland–San Jose, CA	35	Syracuse, NY
		36	Charleston–Huntington, WV
8	Cleveland–Lorain–Akron, OH	37	Kalamazoo–Grand Rapids–Muskegon–Battle Creek, MI
9	Washington, DC		
10	Pittsburgh, PA		
11	St. Louis, MO	38	Louisville, KY
12	Dallas–Fort Worth, TX	39	Oklahoma City, OK
13	Minneapolis–St. Paul, MN	40	Birmingham, AL
14	Baltimore, MD	41	Dayton-Kettering, OH
15	Houston, TX	42	Charlotte, NC
16	Indianapolis–Bloomington, IN	43	Phoenix–Mesa, AZ
17	Cincinnati, OH–Newport, KY	44	Norfolk–Newport News–Portsmouth–Hampton, VA
18	Atlanta, GA		
19	Hartford–New Haven–New Britain–Waterbury, CT	45	San Antonio, TX
		46	Greenville–Spartanburg–Anderson, SC–Asheville, NC
20	Seattle-Tacoma, WA		
21	Miami, FL		
22	Kansas City, MO	47	Greensboro–High Point–Winston-Salem, NC
23	Milwaukee, WI		
24	Buffalo, NY	48	Wichita–Hutchinson, KS
25	Sacramento–Stockton–Modesto, CA	49	Salt Lake City, UT
26	Memphis, TN		

Table 4.1 (*continued*)

A broadcast market encompasses not only the city or cities for which it was named, but also any communities lying within the grade A contour of any stations located there. A look at the top 100 markets as of 1972 shows just how much the cable industry's growth was hampered by the 1969 Report and Order and how much it was helped when the 1972 Report and Order permitted cable in markets 51–100.

50	Wilkes-Barre–Scranton, PA	76	Spokane, WA
51	Little Rock, AR	77	Jackson, MS
52	San Diego, CA	78	Jacksonville, FL
53	Toledo, OH	79	Chattanooga, TN
54	Omaha, NE	80	South Bend–Elkhart, IN
55	Tulsa, OK	81	Albuquerque, NM
56	Orlando–Daytona Beach, FL	82	Fort Wayne–Roanoke, IN
57	Rochester, NY	83	Peoria, IL
58	Harrisburg–Lebanon–Lancaster–York, PA	84	Greenville–Washington–New Bern, NC
59	Texarkana, TX–Shreveport, LA	85	Sioux Falls–Mitchell, SD
60	Mobile, AL–Pensacola, FL	86	Evansville, IN
61	Davenport, IA–Rock Island–Moline, IL	87	Baton Rouge, LA
62	Flint–Bay City–Saginaw, MI	88	Beaumont–Port Arthur, TX
63	Green Bay, WI	89	Duluth, MN–Superior, WI
64	Richmond–Petersburg, VA	90	Wheeling, WV–Steubenville, OH
65	Springfield–Decatur–Champaign–Jacksonville, IL	91	Lincoln–Hastings–Kearney, NE
66	Cedar Rapids–Waterloo, IA	92	Lansing–Onondaga, MI
67	Des Moines–Ames, IA	93	Madison, WI
68	Jacksonville, FL	94	Columbus, GA
69	Cape Girardeau, MO–Paducah, KY–Harrisburg, IL	95	Amarillo, TX
70	Roanoke–Lynchburg, VA	96	Huntsville–Decatur, AL
71	Knoxville, TN	97	Rockford–Freeport, IL
72	Fresno, CA	98	Fargo–Grand Forks–Valley City, ND
73	Raleigh–Durham, NC	99	Monroe, LA–El Dorado, AR
74	Johnstown–Altoona, PA		
75	Portland–Poland Spring, ME	100	Columbia, SC

Source: *Martin H. Seiden*, Cable Television U.S.A.: An Analysis of Government Policy. *New York: Praeger, 1972.*

To comply, the cable operator would need to be aware of which programs and stations the rule would apply to and make appropriate and satisfactory substitutions. By the late 1970s, some common carrier microwave systems would actually offer program substitution as an additional service to their cable system clients.

These anti-leapfrogging and non-duplication rules did not, indeed could not, apply very directly to cable communities lying outside of any broadcast markets. There were a few carriage provisions listed for cable systems in one or more stations' grade B contours, but even these were rather vaguely defined. Otherwise, the provision was simply that cable systems would carry "commercial television broadcast stations that are significantly viewed in the community of the system." Naturally, the most popular and well-funded stations would be the ones "significantly viewed" in a community. So, ironically, America's smallest communities continued to offer the most selective "packages" of cable programming at this time and for several years to come. The same types of small towns where the CATV industry had been founded, and where that industry had flourished during the frozen years of the 1960s, would continue to exemplify the range of channels cable was capable of providing.

Looked at another way, though, the small towns were exempt from what was widely perceived as the main public service benefit of the 1972 Report and Order: upholding the spirit of the 1969 local programming requirement through local origination. The 1972 rules revised the origination requirement so that the main responsibility for operators would be to make channels and facilities available to community residents for public access programming, rather than to carry on the actual creative activity themselves. The well-known concept of "public access" in the United States actually refers to public, educational, and governmental (PEG) programming – non-commercial cable channels set aside for use by members of the community. As in the 1969 rules, the FCC made these local channels a requirement for systems with over 3,500 subscribers. Systems in the top 100 markets had to provide four access channels. This time, though, the expenditures for the operator were intended to be less and the benefit to the public greater.

For this reason, the 1972 Report and Order has become synonymous with the rise of public access facilities. Even though the federal-level public access mandate would be struck down in a 1979 Supreme

Box 4.3   Cable Industry Stocks and the Rise of Paul Kagan

So successful has the cable industry been in recent years that it is hard to imagine a time when that industry was struggling financially. In reality, by the late 1960s several factors had come together to increase operating costs considerably. Equipment had grown both more sophisticated and more expensive. Program origination – which most operators believed would be a long-term obligation – drained revenues. And cities, with no limits on what they could ask in terms of franchise fees, were demanding exorbitant amounts from companies wishing to operate cable systems there. Even after the 1972 rules opened the lucrative broadcast markets somewhat and set a cap on franchise fees, most cable systems were still facing revenue shortages. Smaller, independently owned systems were increasingly being bought out by MSOs, and the MSOs themselves had begun to make public stock offerings – a first for the cable industry.

Industry analyst Paul Kagan, who cites 1974 as the worst year ever for cable finances, found an opportunity in all this. In 1969, he launched Paul Kagan Associates, an industry research firm that would prove key to cable companies' eventual success on Wall Street. Kagan, who had begun his media career in broadcast television news and sports during the early 1960s, had been working at E. F. Hutton when cable's financial crisis hit. His boss had asked him to research the potential of newly public cable companies such as Teleprompter, Cox, and ATC (the company that would form a large portion of today's giant Time-Warner Cable). Kagan reported very favorably on cable, noting its rate of growth, the tax breaks possible to extensive system construction and upgrades, its local monopolies, and the general popularity of television with the American public. Kagan's boss did not see things the same way, however. He was more focused on the bottom line (literally and figuratively), which showed that the heavily leveraged (i.e., in debt to financiers) cable companies were not yet making money.

So Kagan struck out on his own, believing correctly that some investors would be able to see past the short-term debt issue and into the industry's lucrative future – provided they had the right kinds of information. As Kagan explained in a 1999 interview, most potential investors held the same doubts and misconceptions as his boss at E. F. Hutton. "[Cable systems] couldn't make any money the way people knew money could be made. Large corporations were afraid to acquire them or be interested in them because they didn't make a profit, so it would dilute their earnings and they didn't want any part of that. They had no idea that it was going to be 70 million households and how many channels and what a great opportunity it would be" (Kagan oral history). While making the transition from E. F. Hutton to his own research firm, Kagan devised an alternative formula for calculating the value of cable stocks that focused on long-term income potential rather than current dividends. He also attended trade shows and spoke with people in the industry to gather information for potential investors. He published all of this in a newsletter that would grow quickly into something of a bible for those in the industry and those interested in investing in cable.

Kagan's formula revolutionized the young cable industry. As Thomas Southwick describes the breakthrough: "It provided every cable operator in the country with a quick way to calculate his company's net worth. Somebody running a mid-sized system with 2,000 subscribers could look at Kagan's chart, find that the average public company was valued at about $600 per subscriber and go home that night to announce to the family that they were millionaires" (79). The Kagan formula also proved a boon to investors, whether making decisions about buying cable stock or acquiring entire systems.

Kagan's firm eventually began analyzing other media industries, including broadcast television and radio, movies, DBS, and new broadband technologies such as cable telephony. By the 1990s, Paul Kagan Associates was well known for its trade magazines, newsletters, and other research publications. In 2000, the operation was acquired by Primedia, Inc., with Kagan staying on as a consultant. The business is now called Kagan Research LLC (additional information from Kagan Research LLC: www.kagan.com).

Court decision, many communities would retain public access (or initiate it) as part of the local franchising process. And even though public access programming has become the source of hackneyed "bad television" jokes (as epitomized in the 1992 movie *Wayne's World*), there have also been outstanding programs produced over the years.

Public access also remains practically the only way for "ordinary" citizens to bypass the large media conglomerates in getting their messages on television.

More beneficial to the growth of the cable industry itself was the fact that the 1972 rules effectively loosened the ban on distant signal importation. Ostensibly this was meant to create a revenue stream that would support public access facilities, since the imported stations – nearly always big-city independent stations such as Chicago's WGN and New York's WOR – were extremely popular for their schedules of big-league sports, old movies, and television reruns. In effect this represented the beginning of a "slippery slope," however. Even if the American public seemed to idealize an array of educational and other specialty cable programming – as reflected in the Blue Sky writings – when it came right down to it, what most people actually wanted to watch was more of what was already popular on broadcast television. In fact, several of the earliest satellite cable channels *actually were* broadcast stations – called "superstations" – whose signals were beamed up to a satellite transponder and distributed across the continent. Other early cable channels merely resembled broadcast stations, due to their extensive use of off-broadcast syndicated fare. Superstations and cable-only channels resembling them will be discussed more in Chapter 5.

Pay-Cable

Another type of cable channel entering the scene at this stage was the pay-cable channel, heir to the pay-television experiments of the 1950s and 1960s. The eventual defeat of Proposition 15 in California, coupled with the longevity of the International Telemeter trial in Etobicoke, Ontario and the Phonevision trial in Hartford, Connecticut, had sent a positive message about the prospects of pay-television in North America. This was further helped by the FCC's 1968 Report and Order on Pay-Television, which gave both wired and broadcast systems permission to operate on a non-experimental basis.[1] Although there would be a few broadcast-based pay-television systems operating mostly in greater New York and Los Angeles for a brief period in the late 1970s and early 1980s, it quickly became evident that the real future of pay-television lay in cable. In cable subscribers, pay-television entrepreneurs had a market of people already willing to pay for television service. For rural residents, of course this was seen as a necessity in order to enjoy television at all. However the fact that cable increasingly was drawing subscribers from communities already served by broadcast television sent an encouraging message about the potential success of pay channels as part of cable service.

The arrival of pay-cable did not take place without challenges, however. According to the 1968 Report and Order, wire-based pay-television systems were prohibited from using the most popular and readily available movies, those between 2 and 10 years old. They also were forbidden from using any sports events that had been shown on broadcast television during the preceding two years. And they could not show series with "an interconnected plot or substantially the same set of characters," in other words, series-type programming. Furthermore, movies and sports could not make up more than 90 percent of the total programming hours. These restrictions would remain in place until a 1977 Supreme Court case. Nonetheless, the cable industry did manage to become involved with pay-television during the late 1960s and early 1970s.

One of the first and largest pay-cable systems was a movie channel, called Channel 100, that was leased to cable systems by Optical Systems, Inc. of Los Angeles. Channel 100 was a turnkey operation, meaning that once the lease agreement had been signed, the system was ready for operation – fully compatible with existing equipment. This was surely a boon to cable operators who were new to the business of pay-television and its accompanying technology. Channel 100 relied on a system of punched plastic cards, known as "tickets," that could be purchased by cable subscribers individually, as packages, or on a subscription basis. By 1973, four cable systems had adopted the Channel 100 package. Other early pay-cable systems included New York-based TheatreVisioN, tested briefly in Florida; and Home Theater Network (HTN), tested in California. TheatreVisioN, like Channel 100, used plastic tickets – in this case purchased by cable subscribers for a flat monthly fee. HTN relied on a telephone-connected device for ordering. While these pay-per-view type systems seemed promising, none lasted past the early 1970s.

More successful were a group of flat-rate systems, several of which would evolve into today's most long-standing and successful pay-cable networks. During the early 1970s several systems emerged that

Figure 4.2 A promotion for Channel 100 [The Cable Center]

allowed cable operators to program their own pay movie channels using videotapes. These included Gridtronics (owned by entertainment conglomerate and cable MSO Warner Communications), Z Channel (owned jointly by Teleprompter and Hughes Aircraft), and various other systems. Gridtronics is worth mentioning because it eventually grew into The Movie Channel. Z Channel provided the foundation for Showtime.

Even more notable at this stage was the small microwave-based pay channel, Home Box Office, which began in 1972 as a subsidiary of New York City's Sterling Manhattan Cable (a cable MSO partly owned by the Time, Inc. publishing empire). Sterling was one of two cable systems serving Manhattan at the time (the other being Teleprompter). Along with other program origination efforts, the company had begun showing movies on some of its vacant channels. In 1969, Sterling's president, Charles Dolan, presented his board with a proposal for something called the Green Channel – a pay-cable sports and movie service for Manhattan and any other system wishing to pay for it.

When the proposal was finally approved in 1971, the name was changed to Home Box Office. However, due to a franchise provision against pay-television in Manhattan, the channel could not be launched there.

Instead, HBO made its debut on the cable system in Wilkes-Barre, Pennsylvania – transmitted from New York via microwave. The schedule included some feature films, sports events from New York, and other miscellaneous programs (mostly low-budget fare such as coverage of roller derby). Among the first major features for the new pay service was live coverage of the Pennsylvania Polka Festival, featuring "Wally" (Stan Wolowic) & the Polka Chips. It was a low-budget production, though well-suited to its audience in east-central Pennsylvania (an area heavily populated by people of Eastern European descent). Soon other cable systems using regional microwave networks in Pennsylvania and upstate New York were connected to HBO as well, and by its second year HBO had 14 affiliate cable systems and some 8,000 subscribing households. These numbers were promising, especially given that the pay channel's programming possibilities were still subject to the constraints of the 1968 Report and Order. The company was eager to expand beyond the small portion of the Northeast HBO was serving. However most other parts of the country were not so extensively covered by interconnected microwave relays, and for this reason, HBO executives were among those most interested in a demonstration of satellite technology at the 1973 NCTA convention in Anaheim, California. The featured event was a boxing match, transmitted live from Madison Square Garden, and those watching on behalf of Time, Inc. and HBO were wowed. They knew it represented the future of their pay-cable channel.

A federal government policy known as "Open Skies" made it possible for HBO to put its programming up on satellite and ultimately enabled the completion of cable's transformation from a retransmission medium into a true player in the entertainment and information industries. Open Skies was the FCC's plan for open-entry competition in the domestic communications satellite industry. Pressure to implement such a policy had come from the Nixon administration and its newly created Office of Telecommunications Policy (OTP), which was directed by management specialist and engineer Clay T. Whitehead. It was Whitehead's strongly held conviction that competition and industry deregulation represented the fastest, most expeditious path toward maximizing the potential of the new technology.

The State of Cable Technology

At this point it is also worth taking a brief inventory of the state of cable television technology as of the early 1970s. Satellites, of course, would be the major factor in transforming cable from a rural retransmission technology into a multichannel competitor for broadcast television. With a single satellite transponder capable of delivering programming to any receiving dish within its footprint, it would be possible for a start-up cable program service (network) to draw a critical mass of viewers from as great an area as an entire continent (or at least a substantial portion of one). Compare this to the complicated system of AT&T landlines, microwave networks, and bicycled (i.e., physically transported) videotapes that had been needed for the broadcast networks to serve the entire nation via their affiliate stations, and for the affiliates in turn to serve their local areas. For would-be cable networks, the major expenses would be transponder rental and the various costs associated with program production or acquisition.

A lot had changed "on the ground" for cable, as well. Throughout the 1960s signal quality was improved by such innovations as aluminum-shielded, foam-core coaxial cable and solid state amplifier technology. And where CATV operators had been pleased to increase their systems' capacity from three to five channels during the 1950s, it was now possible to offer far more channels – more, in fact, than many operators were yet able to fill. By the mid-1960s a typical CATV system could, in theory, make use of all 12 channels on a VHF television dial. And one invention that was essential in making systems with *over* 12 channels a reality was the heterodyne set-top converter by Ronald Mandell and George Brownstein in 1967. In addition to solving the problem of off-air signals interfering with cable reception, the Mandel-Brownstein converter provided a multichannel tuning device as an alternative to the 12-channel VHF dial. By 1970, the converter was being mass-produced by Oak Communications, Inc. This was no doubt the impetus for the FCC's making 20-channel capacity mandatory for systems operating in broadcast markets as part of the 1972 rules. Clearly the community antenna era was over.

The International Scene

It should be pointed out that during this period, the early signs that CATV was being transformed into multichannel cable television were apparent almost exclusively to North Americans. Some nations in Western Europe were starting to allow second public service channels, and in a very few cases commercial networks – suggesting an appetite for programming variety that would only grow in the years to come. But elsewhere in the world, television itself was still a new phenomenon, with the vast majority of people completely unable to afford access. Satellites also lay in the future for most nations.

Yet these years also mark a point when people in the United States clearly were beginning to question the sort of television industry and regulatory framework that had developed here – and began to look beyond US borders for alternatives. No doubt some of those individuals and groups who perceived cable as the antidote to unfulfilled expectations for broadcast television during this period had seen what television had to offer in countries where the public service model dominated. Some evidence of this lies in the fact that the newly formed PBS became known very early on for its use of imported programs produced by and for the BBC.

In the case of Canada, where its CATV industry had been developing in somewhat of a symbiotic relationship with that of the United States *up to this point*, it started to follow a different path in the 1970s. Canada's cable industry still existed first and foremost to deliver American television to Canadian households, but protectionist policy-makers were starting to take notice and voice objections to this practice. In 1968 a Broadcasting Act transformed the Broadcast Board of Governors, a fairly permissive regulatory authority, into the Canadian Radio-Television Commission (CRTC), which over the coming decades would create and enforce strict "Canadian content" quotas for both ownership and content in that nation's broadcast and cable television industries. This represents one of the earliest attempts to limit the export of US culture via television – something that eventually would concern other nations as they worked to develop their own media infrastructures and as new media increasingly enabled the free flow of information across national borders. Compared to other nations, particularly those in the developing world, Canada has been reasonably successful in these efforts, allowing the creation of some critically acclaimed programs by the CBC as well as private networks (some of which have been exported to the United States with good response, including the *Anne of Green Gables* movies and more recently *The Red Green Show* and *Trailer Park Boys*).

Interestingly, Canada also set some important examples for US regulators in its efforts to foster indigenous programming – particularly with regard to public access cable programming. During the late 1960s an organization called Challenge for Change, an outgrowth of the National Film Board of Canada, began using Sony Portapaks in its campaign to eradicate poverty by documenting its causes. As Laura Linder explains:

> In 1970, the prototype for public access television was developed in Thunder Bay, Ontario. A local civic organization initiated a plan to operate a cable channel for the local community, relying on Challenge for Change to supply the equipment and train members of the community in its use. The group eventually cablecast locally produced community programming four hours a day. Shows were produced live as well as taped, and the important precedent of using half-inch video was established (half-inch video was not considered broadcast quality but was deemed acceptable for cablecasting). The Thunder Bay Community Program lasted less than a year, but it demonstrated the potential for public access television. (4)

A subsequent 1971 CRTC Policy Statement on Cable Television required public access as part of cable's development, and US policy-makers can be credited with paying attention to their northern neighbors – even if the public-spirited policy measure that ensued would not be a lasting one.

Clearly the late 1960s and early 1970s were years of trial and error for the US cable industry and its government overseers. While many would cite the late 1970s and the 1980s as the pivotal period in cable history, due to the first uses of satellites for transmission of cable programming, such a claim warrants further consideration. While the years following 1975 represent a time when North Americans had their first look at the cable of the future, nearly everything that allowed that scenario to come about took place during the confusion of the late 1960s and early 1970s: regulations requiring the development of non broadcast-derived cable programming combined with other regulations permitting more and more broadcast-derived programming, a plethora of plans to guide the development of satellite-served cable combined with a regulatory philosophy that deregulated the satellite industry, and easing of concerns surrounding copyright protection combined with an increasing array of cable-based systems for delivering copyrighted material. In the next chapter we will see evidence of all of this "coming home to roost," so to speak, as the cable industry grew less regulated almost by the day and increasingly relied on the programming and precedents of broadcast television to enter new markets.

NOTE

1   Even at this stage there were challenges. In September 1969, the US Circuit Court of Appeals for the District of Columbia ruled against pay-television complaints filed by the National Association of Theater Owners and the Joint Committee against Toll TV – affirming the FCC's authority to license pay-television systems.

FURTHER READING

Easton, Ken. *Building an Industry: A History of Cable Television and its Development in Canada*. East Lawrencetown, Nova Scotia: Pottersfield, 2000.

Howard, H. H. and S. L. Carroll. *Subscription Television: History, Current Status, and Economic Projections*. Knoxville: University of Tennessee, 1980.

Linder, Laura R. *Public Access Television: America's Electronic Soapbox*. Westport, CT: Praeger, 1999.

Streeter, Tom. "Blue Skies and Strange Bedfellows: The Discourse of Cable Television," in *The Revolution Wasn't Televised: Sixties Television and Social Conflict*, ed. Lynn Spigel and Michael Curtin, New York: Routledge, 1997, 221–42.

Chapter 5

Cable Meets Satellite: 1975–80

In 1965, the US cable industry had very few means by which to grow – banned from the top television (and population) markets, waiting for pay-television to gain FCC approval, waiting for access to communication satellites, and fighting for the right to use copyrighted program material. A decade later, nearly everything was different. The cable industry had the go-ahead (at least provisionally) for practically every new business venture it wished to try. The 1975 satellite debut of fledgling pay-cable service Home Box Office marked the definitive arrival of cable television's modern era. From 1975 to 1980 the most dramatic events in the cable industry were in the areas of deregulation and technological innovation that would help bring new programming outlets to the television audience. By 1980, there would be enough satellite-carried cable channels to suggest cable's potential to compete with broadcast television, perhaps even to offer the niche services unknown to the "big three" broadcast network oligopoly of the time. However, also by the late 1970s visions for cable as a public service medium had effectively given way to the demands of a very lightly regulated television marketplace. With no clear plan for the growth of satellite cable, the first players to emerge were those with the financing to rent satellite transponders, existing programming infrastructures, and a ready appeal to potential audiences.

Cable Regulations

The handful of cable networks launched in the late 1970s were helped by an increasingly favorable policy climate. Satellite uplinking costs and transponder rental still kept most would-be cable networks out of the competition. And many local operators found the receiving dishes and converter boxes prohibitively expensive when trying to offer the new services to their subscribers. But the Open Skies policy had at least eliminated most of the legal obstacles to satellite start-up operations. From the cable industry's perspective, things were looking good by this point. Social reformers, of course, had much less to celebrate as it became increasingly clear that cable would develop as a commercial medium with relatively little to offer that did not duplicate broadcast television. Other policy developments in this later half of the decade would only provide more encouragement to a commercial cable industry.

The ink had no sooner dried on the FCC's 1972 Report and Order than its regulatory provisions began to be overturned. The overall climate in Washington was deregulatory at this point. The FCC, under pressure from the White House (first Nixon and then Ford) and Congress (notably in a 1976 report by the House Communications Subcommittee criticizing the 1972 rules), began to review its policies. What ensued was a series of regulatory revisions and court cases that, by the end of the decade, would eliminate virtually every factor limiting the cable industry's growth.

In 1974, following a challenge in federal court, the FCC abandoned the rule requiring cable systems serving over 3,500 subscribers to originate programming (ironically, the FCC had won in this case, but chose to concede anyway). In 1976, the agency permitted an unlimited number of distant signals to be imported by a cable system once local channels were off the air. Then in 1980 it eliminated the remainder of the "antileapfrogging" rules, allowing operators to import signals from anywhere in the country (recall that the 1972 rules had set regional parameters for distant signal importation). This naturally gave a huge boost to those programming entrepreneurs involved in relaying the signals of popular independent broadcast stations via microwave or satellite – as will be discussed later in this chapter.

While all of this was taking place, a groundbreaking case involving cable programming, *Home Box Office v. FCC*, was moving through the court system. In 1973, HBO – keenly aware of how the FCC's 1968 Report and Order on Pay Television affected its potential to succeed in the competitive television programming marketplace – appealed to the Commission to repeal the ruling. While the Report and Order writ large had made it possible for pay-cable operations such as HBO to exist at all, the portion of it known as the "antisiphon-ing" provisions that prevented pay services from using movies between two and ten years old was stifling. Hearings on this matter would be difficult and protracted, however. Both movie theater owners and television broadcasters weighed in to oppose lifting the restrictions; in fact most wanted those restrictions to be tightened even further. The term "siphoning" was provocative to established entertainment industries for various reasons. If the restrictions on pay-cable were lifted, movies would move more quickly out of theaters and onto television (which indeed has happened in more recent decades), but it would take longer for those same movies to reach commercial broad-cast networks (which also has happened). As National Association of Broadcasters (NAB) president Vincent Wasilewski argued in his testi-mony, "If cable interests don't intend to siphon, why are they so intent upon removing present antisiphoning rules?" (cited in Hard decision). For their part, the Hollywood studios, represented at the hearings by Motion Picture Association of America (MPAA) pres-ident Jack Valenti, favored lifting the rules. The studios had been in a financial slump for about a decade and pay-cable stood as a promising new exhibition outlet and revenue source.

In 1975, the FCC issued a revised set of the rules, a measure intended as a compromise among the competing interests. These failed to satisfy any of the parties concerned, however, and the matter was appealed to the federal courts. In 1977, the US Court of Appeals for the District of Columbia ruled that the FCC had exceeded its regulatory authority with the antisiphoning rules, thereby allowing HBO and other cable networks unrestricted access to movies and television series not previously shown on broadcast television. Replac-ing the broadcast networks as Hollywood's first television exhibition window would be the critical factor in the future success of HBO and other pay-cable networks.

Still another boon to HBO and other pay-cable services was the FCC's 1979 repeal of strict technological specifications and licensing requirement for satellite receiving dishes (also known as "receive-only," or "TVRO" dishes). Until this point HBO was available almost exclusively to subscribers of MSO-owned systems, as they generally represented the largest and most profitable systems and also enjoyed bulk discounts from equipment manufacturers. After the 1979 repeal, the cost to a local cable operator immediately dropped to a level that many more could afford: from $100,000 to around $15,000–$20,000 per dish – which in turn further lowered the cost of the dishes. Of course, by this point HBO was hardly the only network available via the receiving dishes – making the significantly lowered expenditure even more justifiable to operators.

In 1980, the syndicated exclusivity provisions of the 1972 Report and Order were lifted, making programming more accessible to both recently launched and future cable networks. This was primarily the result of two studies released by the FCC in 1979, the Syndicated Exclusivity Report and the Economic Inquiry Report. The studies had investigated the extent to which competition from cable (whether in the form of distant signals or satellite channels) stood to harm local broadcasters and therefore threaten the public interest. The two reports concluded essentially the same thing: cable television did not have a demonstrably negative financial impact on local broadcasting. Rather, it constituted, at most, a minor threat to local broadcasters' financial health and no threat to the public interest (Parsons and Frieden: 56). The controversy did not go away, however, and another version of the syndicated exclusivity rules would be implemented in 1988.

But as of 1980, the only regulations limiting cable's expansion were the sports-related "blackout" provisions imposed by the FCC in 1975. These rules were aimed at maintaining a level playing field between cable systems and broadcasters with regard to televised local sports events. Due to the possibility of cutting into the ticket sales of sports franchises, broadcasters were severely limited in their ability to carry local sports events. The ability of local cable systems to import distant signals brought with it the possibility of sports events being shown in their home markets in spite of the broadcast prohibition. Thus the FCC prohibited any cable system within a 35-mile radius of any broadcast station blacked out for a sports event from importing that event on a distant signal (or, later, on a regional or national satellite network). These complex rules remain in place today, and have been extended to include direct-to-home satellite as well.

Copyright

Meanwhile, the long awaited copyright legislation had finally arrived in 1976. The cable industry had spent many years in court battling copyright cases: first in *Fortnightly v. United Artists* and then in *Teleprompter v. CBS*. Both cases and their outcomes represented tentative victories that lent the industry some optimism in its pursuit of new revenue streams. Nevertheless, there still had been no definitive copyright *law* to guide cable (or indeed any electronic media industry). This finally changed with passage of a new Copyright Act, the first since 1909. Of course, all the time Congress had spent deliberating and conducting hearings on how the legislation should be written represented years in which the cable industry was anything but stagnant. With complete uncertainty as to what lay in cable's future, the final version of the Act could, at best, address the shape and scope of the industry only as it existed at that time. And of course cable was only one of the many entertainment and information industries implicated in copyright issues.

When the 1976 Act was complete, it represented something of a juggling act. Writers Thomas F. Baldwin and D. Stevens McVoy offer a clear explanation of the competing issues it sought to address:

> The first purpose is "to maximize the availability of creative works to the public." The Copyright Act grants monopoly property rights in the production and publication of literary, musical, artistic, and dramatic works. It is assumed that these rights are necessary incentives to creators. But also, the ability to avail oneself of creative work is a public benefit that must be facilitated. Copyright law provides an administrative procedure by which original material may be shared.
>
> The second purpose is "to afford the copyright owner a fair return for his creative work and the copyright user a fair income under existing economic conditions." The Copyright Act attempts to protect the livelihood of the copyright owner and at the same time offer fair incentives to the user of copyrighted materials (such as a cable company) so that they may be disseminated fully.
>
> A third purpose is "to reflect the relative roles of the copyright user in the product made available to the public with respect to relative creative contribution, technological contribution, capital investment, cost, risk, and contribution to the opening of new markets for creative expression and media for their communications." This acknowledges the complexities of the modern creative process and the role and costs of communication technology in making original work available.
>
> The final purpose is "to minimize any disruptive impact on the structure of the industries involved and on generally prevailing industry practices." The implementation of the copyright law is a delicate task.

> As new communication technologies emerge and royalty rates are estab-
> lished and adjusted, the Copyright Royalty Tribunal must be cautious
> not to upset or restructure communications industries to the detriment
> of copyright owners, disseminators, or the public. (186–7)

Complicated though it was, the Act finally gave cable operators
and others the assurance of a systematic payment mechanism that
would allow them reasonable and undisputed access to television
programming.

In the case of cable, copyright payments are now settled through a
"compulsory license" system that allows operators to retransmit local
signals without paying royalties but which requires a flat payment for
programming on distant signals. The compulsory license system is
administered by the Copyright Royalty Tribunal (five commissioners,
each appointed by the US President for a term of seven years). The
Tribunal collects fees from cable operators (based on their subscriber
revenues) and then redistributes that money among copyright holders. This not only saves the cable operators money (while still providing fair compensation for the material they use), it also eliminates a
great deal of bookkeeping. The 1976 Copyright Act is still in place,
amended to address new technologies and practices.[1]

Home Box Office: The First Satellite Cable Network

The September 1975 satellite debut and successful start-up years of the
pay-cable service Home Box Office represent the cumulative impact
of the policy shifts discussed so far in this chapter. HBO not only
heralded the beginning of satellite-served cable networks; it also initiated a set of precedents for how cable, as a new programming medium,
would hold its own alongside the broadcast television behemoth.
During its brief microwave phase, HBO had already established the
benefits of a subscription model over a pay-per-view model. This
would prove crucial in satellite cable's early years, since cable subscribers were more likely to bear with programming flops (of which
there were many) if they were not paying for them directly. HBO,
despite legendary programming coups (especially its live coverage of
major boxing matches), had to fill most of its early schedule with
low-cost programs such as minor sports and obscure movies. Other

Figure 5.1   Gerald R. Levin [The Cable Center]

early satellite networks would, of course, face similar programming challenges.

HBO's launch brought with it another new facet of cable operations: marketing. Until this point, cable operators had not needed extensive marketing efforts to convince people to sign on; after all, in most places it had been the means to any television service at all. With pay-cable, however, subscribers had to be persuaded to double their monthly payments for a completely new level of service. This too only prefigured future challenges the industry would face – especially as it tried to convince residents of major broadcast markets that cable subscriptions would bring them something worth paying for in addition to the broadcast channels they could receive at no direct cost.

HBO's pioneering shift from a terrestrial-based regional service to one delivered nationally (and eventually internationally) via satellite was not an easy one, since the cable industry itself was not fully convinced it would be feasible. Parent company Time Inc. had even contemplated pulling the plug on the HBO experiment – which had been losing money from the start. In 1973, Time had spun off the New York City and Long Island cable systems that had been part of Sterling Cable. In doing so, it also let go of Charles Dolan, HBO's founder.[2] But in the meantime Gerald Levin, a young attorney hired by Dolan in 1971, had become president of HBO and managed to persuade the parent company to keep it operating. For Levin, this was the beginning of a highly successful media career, and his early accomplishments with HBO are telling of this potential.

As discussed in the previous chapter, three cable-related companies cooperatively had held a demonstration of satellite technology (in which they all held a financial stake) at the 1973 NCTA annual convention in Anaheim: HBO, representing the industry vanguard in programming; Teleprompter, the largest MSO at the time; and Scientific Atlanta, a major cable equipment supplier. The satellite telecast was of a boxing match live from Madison Square Garden. It represented many things for the shifting industry – including the multiple businesses and technologies that were coming together under the rubric of something called "cable television" and that newly redefined medium's ability to address areas not fully covered by broadcast television. It might not have been exactly what the Blue Sky planners had envisioned just a few years earlier, but it definitely was something new.

HBO made its first satellite telecast, another boxing match, to actual subscribers in September 1975. This bout, Ali vs. Frazier live from the Philippines ("The Thrilla in Manila") helped establish the pay-cable service as a home for televised boxing – a programming niche vital to its success during the tentative start-up years. Of course, Irving Kahn, who was instrumental in getting a satellite-served HBO off the ground, had foreseen cable's promise with boxing nearly two decades earlier with his experimental closed-circuit pay-per-view events.

During the later 1970s – years before other cable "brands" such as CNN and MTV had even been launched – HBO was establishing cable programming precedents such as theme days (e.g., "Gasless Saturdays and Sundays," free promotion days that played off the oil crisis and people's choice to stay home and avoid unnecessary driving) and the airing of programs (such as BBC productions) not selected by the major broadcast networks due to uncertain audience appeal. At first HBO's programming experiments proved something of a drain to the financial resources of Time Inc., as the wealthy publishing company had to foot the bill for various programming ventures that either flopped completely or at least failed to recover their initial investment. But over the next two decades HBO's established programming success would serve as a model for other entertainment corporations, particular in the synergy-driven frenzy of mergers and acquisitions that would begin during the late 1980s. Indeed, one of HBO's most enduring and imitated program-acquisition strategies was introduced in 1977, just after its court victory: serving as a first television distribution window for Hollywood movies – with rights often secured by Time Inc. and HBO through upfront co-financing deals. One such deal was initiated with Columbia Pictures at the early date of June 1976. This deal gave HBO access to films such as *Taxi Driver* (1976), *Fun with Dick and Jane* (1977), and *Close Encounters of the Third Kind* (1977) (HBO makes deals).

In fact, the Hollywood movie studios were paying close attention to HBO from the start, particularly its movie co-financing deals. They reasoned that if an outfit owned by a publishing company could be profiting from their product, it would make sense for them simply to start their own pay-cable network. It was to be called Premiere. The Premiere venture was the joint effort of Columbia, Paramount, Universal/MGM, and Twentieth Century-Fox, along with the Getty Oil Corporation. While Premiere's founding concept – a nine-month exclusivity window for films produced by the partner studios – echoed HBO's strategy, and while the studios (plus Getty) had more financial resources than most start-up satellite networks, Premiere's launch was ultimately blocked by the US Justice Department on antitrust grounds (a precedent dating back to the 1948 Paramount Decree). Had this not happened, pay-cable might have developed very differently, since Premiere's partner studios surely would made it difficult if not impossible for HBO and other pay-cable outfits to gain access to their movies.

By the time of the Premiere debacle in 1979–81, some other pay-cable competition had emerged for HBO, coming not from Hollywood but from some of the cable industry's very earliest forays into pay-television. One was the Movie Channel, which made its satellite debut in 1979. The Movie Channel had developed out of a videotape-based movie delivery system, known as Gridtronics, with its origins in a late 1960s experiment by cable MSO Television Communications Corporation (TVC). It did not launch on cable until 1973, however, following its purchase by Warner Communications (which naturally helped the start-up pay service to acquire movies). During the years of its terrestrially-based operations, Gridtronics was known as Warner Star Channel.

Also joining HBO in its early years – in fact in direct competition with it – was Showtime. Showtime began in 1976 as a microwave- and videotape-based pay service serving cable systems in New York City and on Long Island. It targeted local systems not willing or able to purchase the satellite receiving dishes needed for HBO. Two years later it became evident that satellite delivery was the key to pay-cable success, and Showtime also uplinked its programming. Competition between these two satellite pay services became even stiffer at that point, with both available nationwide. The only significant difference in their programming was that HBO had begun to produce some of its own programming while Showtime acquired all of its early pro- gramming from outside sources. Over time, this situation would change. The two networks continue to be the main pay-cable (and direct broadcast satellite) rivals – with both offering big-ticket sports, recent Hollywood movies, and award-winning original movies and series.

Meanwhile, other competition had emerged in the form of *basic cable networks*. Recall that the difference between pay-cable networks such as HBO and Showtime and basic cable networks lies in the way cable subscribers pay for them: directly or *à la carte* in the case of pay-cable and as part of a *bundled* package in the case of basic cable. Even in the early twenty-first century it is not possible for subscribers to choose specific basic cable networks and pay for them directly; instead they must pay for a number of basic networks, including (probably) quite a few they never watch.

Basic Cable: The Superstation

Pay-cable outfits had been the first to perceive the benefits of distri- buting their programming by satellite. After all, they already had begun to explore various programming sources and had some ideas about scheduling. Having a business and programming infrastructure in place before moving to satellite was important since dealing with the intri- cacies of the new technology was challenging enough. Given this logic, then, it should be no surprise that the second category of satellite cable channels to make their debut were the superstations. A cable sup- erstation was – and continues to be – nothing more than a local broad- cast television station whose signal has been uplinked to a satellite transponder for transmission to areas beyond its local or regional market.

The superstation concept actually dates back to the importing of big-city independent stations (by either antenna or microwave) by CATV operators during the 1950s and 1960s. However the first (and still the only) station owner to deliberately make a signal available by satellite was media mogul Ted Turner, with his Atlanta station WTCG (later renamed WTBS), in 1976. Turner had inherited his father's billboard company during the late 1960s, but quickly grew bored with the steady, reliable revenue stream it brought. Turner – perhaps the media industry's all-time most outspoken and adventuresome personality – wanted to take more business risks. And at that time cable television programming was an extraordinarily risky venture.

It is often said that a major coup for Turner and his superstation was the notion of counterprogramming the broadcast networks' evening newscasts with *Star Trek* reruns. He guessed correctly that many Americans had lost interest in the news – an intuition that also, ironically, would lead him to found the Cable News Network (CNN) a few years later. In fact the WTBS schedule included a number of reruns. The station was also known for its classic movies, a genre near and dear to the notoriously nostalgic and romantic Turner. His love of old movies would manifest itself again in the 1980s and 1990s with the founding of TNT, Turner Movie Classics, and other networks specializing in classic and original movies. But probably the most audacious programming move for Turner in getting the superstation off the ground was his outright purchase of major sports teams – first the Atlanta Braves (baseball) and then the Hawks (basketball) and Thrashers (hockey) – thus guaranteeing television rights to all home games. Because of this, Atlanta's professional sports teams began to draw fans from as far away as Alaska.

Other local independent stations also attained superstation status during the late 1970s: New York's WOR (1979) and Chicago's WGN (1978).[3] These represent a different category of superstation, however. TBS is considered an "active" superstation due to Turner's own efforts to uplink the signal. The other superstations are considered "passive" because their signals have been uplinked by third parties (such as Eastern Microwave and United Video) without their owners' consent (a practice that later would be subject to a complicated set of regulations). Like WTBS, the passive superstations featured schedules filled with reruns, old movies, and major-league sports. Still, unlike

Turner, their owners did not actively pursue advertisers outside their local markets – at least not in the early years.

Basic Cable: Other Networks

Since the superstation stood as a ready-made cable channel, with established audience appeal, it is fairly easy to understand why so many of the other early basic cable networks bore a strong resemblance to broadcast television – even if they were not *literally* broadcast stations. It cannot be overstated how risky a venture satellite cablecasting was at this stage. Financing transponder rental at a cost of well over a million dollars per year was just one of the challenges. Acquiring or producing affordable programming was another. And maintaining a reliable revenue stream – often though not always through advertising – was a third. So a rather eclectic mix of satellite channels emerged during the late 1970s – no doubt perturbing many of the previous decade's Blue Sky visionaries.

The first of these was the Christian Broadcasting Network's cable service (CBN-C), launched by televangelist Pat Robertson in 1977. CBN had been operating as a terrestrial broadcast network since 1961, but Robertson saw satellite carriage as an efficient alternative to the clumsy bicycling (i.e., physical transporting) of videotapes to affiliate stations. The centerpiece of Robertson's media holdings was his religious news and talk show, *The 700 Club*. He envisioned satellite cable as a means of distributing this program, along with a schedule of other religious-themed programs, to his television flock. But Robertson was presented with a challenge: how to fill the entire programming day that was suddenly available with the satellite. The broadcast affiliate stations had incorporated CBN programs into programming schedules that otherwise were determined at the local level – much as affiliates of the major broadcast networks combine network programming with local news, syndicated programming, and so on. The satellite programming day belonged entirely to Robertson and CBN, though.

Robertson, taking a cue from his broadcast affiliates, began to purchase the rights to old movies and television reruns. His hope was to gradually replace the syndicated fare with original CBN programs as the network recovered its start-up costs. But it turned out that the

Figure 5.2   Marion G. "Pat" Robertson, in front of a satellite receiving dish
[The Cable Center]

so-called "filler" was very popular with audiences – drawing traditional couch potatoes along with the faithful. CBN retained *The 700 Club* and introduced some other religious programs, but it also became the TV "home" to syndicated evergreens such as *Bonanza* and *The Rifleman*. Robertson had learned the lesson Turner already knew: Americans' appetite for the old and familiar is insatiable. This would be critical to the network's fortunes during the next two decades, as CBN-C gradually shifted from a religious orientation to a more generalized "family-friendly" identity. By the time CBN-Cable changed its name to the Family Channel in 1988, relatively few of its viewers considered it a specifically religious channel. By the early 1990s the

Family Channel was one of the most popular basic cable networks in the United States and, when sold to Rupert Murdoch's News Corporation and children's television producer Saban Entertainment in 1998, commanded a price of $1.9 billion.

PTL, the satellite network launched by televangelists Jim and Tammy Faye Bakker in 1978, followed a somewhat different path. Like Robertson, the Bakkers had been able to draw on donations from their television congregation when starting their cable network – a revenue stream not enjoyed by other satellite cable programming entrepreneurs. They, however, chose to fill their programming day with original talk shows, children's programs, religious-themed soap operas, and other inexpensive studio-produced fare. A lesson can be taken from a comparison between this strategy and that of CBN. The Bakkers, already enmeshed in scandal, lost control of PTL during the late 1980s. New owners renamed it the Inspirational Network and broadened its programming array and target audience considerably.

The existing advertising revenues of an Atlanta broadcast station helped Turner get his cable operation off the ground, and contributions from television congregations helped the televangelists. Then, in 1979, another early satellite cable network was launched successfully due to contributions from the cable industry itself. C-SPAN, the Cable-Satellite Public Affairs Network, represented a public service gesture on the part of cable operators – and has functioned in this capacity ever since. The idea for C-SPAN came from Brian Lamb, who had worked for President Nixon's OTP (discussed in Chapter 4) and subsequently served as *Cablevision* magazine's Washington bureau chief. Lamb's interest in both the cable industry and national politics led him to propose a cable network that would offer continuous coverage of the US House of Representatives. He was supported in this effort by House Speaker Thomas P. "Tip" O'Neill (D–Mass.) and Communications Subcommittee Chair Lionel Van Deerlin (D–Calif.), both of whom advocated allowing television cameras into the House chambers. Lamb had more difficulty convincing cable operators to support the effort, however. Their budgets already were stretched from the need to purchase new technology such as satellite receiving dishes and updated converter boxes. In addition, many of them also had invested in (or were thinking of investing in) other start-up networks. They did not feel ready to gamble on a public service initiative.

Box 5.1   Women in Cable Television

Among the cable pioneers discussed in earlier chapters of this book, few women stand out. In fact the CATV industry was about as male-dominated as any industry could be – ironic, considering that so many local CATV systems have been characterized as "mom 'n' pop" operations. There were some fairly standard roles for women who were in business with their husbands: bookkeeping, office management, secretarial work, and so on. These CATV "moms" seldom took part in the trade associations or otherwise interacted with other industry professionals. The NCTA even went so far as to schedule flower shows and shopping excursions for women accompanying their husbands to the conventions (as reported in *NCTA Membership Bulletin*, the organization's newsletter at the time).

For this reason, women like Yolanda Barco and Polly Dunn stand out. As discussed in Chapter 4 (see Box 4.2), Barco and her father were both law partners and co-owners of the Meadville (Pennsylvania) Master Antenna system. Barco's legal and business expertise were in high demand by the young CATV industry. Archival photographs from her early career almost always portray her as the sole woman among groups of men in the industry. Dunn, a cable operator in partnership with her husband, hardly fit the stereotypical "mom" role, either. She was instrumental in forming regional trade associations in the South and became active in the NCTA, serving on its board of directors and advocating for the interests of independent systems. Barco and Dunn are virtually the only two women recognized among CATV's earliest pioneers.

By the 1980s, however, many more women had joined the ranks of cable industry professionals – including such powerful figures as Geraldine Laybourne, Kay Koplowitz, June Travis, and Beverly Hermann. A combination of factors had drawn more women into the cable industry – including the successes of the Women's Movement, the increase in numbers of women in media industries generally, and the opening of new cable-related professions in such areas as programming and marketing. Laybourne, for example, had extended her own love of television as a child into her founding role in the children's network Nickelodeon.

The increase in women working in various cable-related professions led to the formation, in 1979, of the organization Women in

Cable Television (WICT) – which has grown into a major industry trade organization. In its nearly three decades, WICT has acted as an advocacy organization for women's advancement in the cable industry. WICT offers networking and informational forums, professional development seminars, and a career service. It also recognizes women's leadership through its business, technology, and programming honors, as well as Woman of the Year and Woman to Watch awards. The Washington, DC-based organization currently boasts over 5,000 members from an extensive range of cable-related professions, and is represented by local and regional chapters across the United States.

Then Robert Rosencrans stepped in to help Lamb. Rosencrans, the CEO of cable MSO UA/Columbia Cablevision (whom Lamb once had interviewed for *Cablevision*), had a longstanding interest in cable programming. He had launched the Madison Square Garden sports network (MSG) as a regional terrestrial service in 1969 and uplinked it to satellite in 1977. The satellite that delivered MSG also carried UA/Columbia's children's service, Calliope, which programmed a few hours per day. It was only after Rosencrans agreed to donate some of the less popular daytime hours on MSG for use by C-SPAN that other operators began to come on board, contributing a cumulative sum of $25,000 for C-SPAN's launch. Support for this commercial-free basic cable network has continued in the form of outright donations as well as per-subscriber fees.

To fill hours when the House was not in session, Lamb began to televise committee hearings, Press Club presentations, and, eventually his own interviews. The latter, including his program *Booknotes*, have become popular C-SPAN offerings. C-SPAN (and C-SPAN 2, launched in 1986 to cover proceedings of the US Senate) probably come closest of all cable channels to meeting the public service goals articulated by so many of the Blue Sky visionaries.

Not surprisingly, it has been a struggle for the C-SPAN networks – which have not deviated from Lamb's original policy of continuous live and unedited coverage of the legislature – to compete for *shelf space* (i.e., inclusion in cable operators' channel selections). Although Ted Turner's Cable News Network (discussed in the next chapter)

Figure 5.3   Brian Lamb [C-SPAN]

has become synonymous with continuous cable news coverage, C-SPAN and C-SPAN 2 have also contributed to people's expectations for news in a multichannel world. While commercial television news outlets (including CNN as well as the broadcast networks) provide "sound bites" that are easily absorbed and retained by audiences, the C-SPAN approach assures us that more in-depth information is available should we want it.

C-SPAN, although it did set new expectations for television news, nonetheless was operating in a programming area already familiar to the television audience. The same was true of the cable sports networks that launched in the 1970s: MSG (Madison Square Garden Network) and ESPN (Entertainment and Sports Programming Network). MSG, with its established presence on cable, ironically would lose its sports focus within the first few years of satellite operations, becoming the superstations' lookalike network USA in 1980. ESPN would be the network that helped the cable industry to set new expectations for televised sports – though this did not happen overnight.

ESPN got off the ground with a large investment from Getty Oil (controlling 85 percent of the network's stock), but it was still a rough start. The way journalist Michael Freeman describes ESPN's September 1979 debut might have described any of the early satellite networks. It was far from glitzy and glamorous:

> Gaping holes still marred the foundation [of the network's newly acquired building in Bristol, Connecticut] and bulldozers continued to move mounds of earth. Incredibly, ESPN was going to do its first show with the studio only three-quarters built. Inside, the set still had not arrived, wires dangled from the ceiling, mosquitoes buzzed through the building, and a small family of skunks wandered an abandoned corridor . . .
> [F]irst, at 7 A.M., September 8, there was munster hurling. Two million Americans had ESPN on their cable systems, and probably only a handful of them knew what they were watching early this morning as men wearing short skirts ran around a field playing a version of field hockey. It was the type of odd alternative programming that would define ESPN for years. (91–3)

Like HBO and other predecessors – especially those programming sports – ESPN faced the almost insurmountable challenge of offering material viewers would find interesting and worthwhile, even when the television rights to most major sports events had long been in the hands of broadcast networks. Today, when an infinitely wealthier ESPN is virtually synonymous with televised sports, it is hard to imagine the network getting by with events that would appeal to only the most diehard or eclectic of sports fans.

One of ESPN's early triumphs involved its inroads in sports news reporting. The popular *SportsCenter* program was born simultaneously with the network itself. As Freeman recounts, the affable *SportsCenter* host, Chris Berman, quickly made a name for himself and the program by giving funny nicknames to some of the athletes mentioned in otherwise dry sports recaps. Other successes included obtaining the rights to various NCAA events as well as Canadian Football League games. Furthermore, ESPN's 24-hour programming day allowed it to schedule much more sports programming than the broadcast networks or general-interest cable networks, so even if it could not air some premiere events, it did have success with otherwise unaired second-tier events – events that, in some cases, proved to be nearly as popular.

Another programming niche that cable inherited from broadcast television was Spanish-language programming. SIN, the Spanish International Network, had its beginnings in the collection of stations that, since the early 1960s, had been sharing programs imported from Latin American countries (*telenovelas*, sports, movies, talk shows, etc.). Just over a decade later, enough stations had affiliated with this network that it would become cost-effective to share programming via satellite once that technology became available. What began as satellite distribution of programs to stations (similar to what the major broadcast networks were doing by this stage) eventually also became a source of programming for cable systems. By the mid-1980s, SIN was functioning informally as both a broadcast and a cable network. Meanwhile, SIN had also launched the full-fledged satellite cable network Galavision in 1979 to make its programming even more widely available.

Yet another popular basic cable network that started in the 1970s was Nickelodeon, a network whose origins resembled those of the pay-cable networks more than other basic networks. Although Nickelodeon did program in an established and popular television genre, children's programming, it had no pre-cable existence – merely a pre-satellite existence. Nickelodeon actually began in 1977 as part of Warner Communications' experimental QUBE interactive cable system, tested in Columbus, Ohio. QUBE had been the cable industry's most direct response to Blue Sky visions for cable. The system offered movies as well as local news, sports, and weather, but its highlight was interactive programming – initially at-home polling on local issues, consumer information, and other programs that took advantage of the five response buttons in the "selector box." The most popular category of interactive programming ultimately proved to be game shows with cash prizes.

Even these drew only a small percentage of the available audience, however, and the system was phased out in the early 1980s. The children's component of QUBE, known as Pinwheel at the time, was the QUBE feature that survived the demise of the interactive system. Pinwheel's mainstay was the 12-hour *Pinwheel House*, an amalgamated program similar in format to *Sesame Street*. Parents at the time were no doubt pleased to have something for their children to watch other than the commercially driven cartoons that were on the rise or the hosted studio shows whose heyday had long since

Box 5.2 The Remarkable Rise of Spanish-language
Television in the United States

With Latinos and Latinas representing the fastest growing minority group in the United States today, it is easy to forget that
Spanish-speaking people have occupied the territory that now
comprises this country since the sixteenth century. This is particularly true in the Southwest, where Spanish-language media –
first newspapers and then movies, radio, and television – have
flourished through many generations. It is not surprising, then,
that there is a long history behind the various Spanish-language
and Hispanic-targeted English-language cable/satellite channels
available today.

The availability of Spanish-language radio in the United States
dates back to at least the 1930s, when Mexico's Azcárraga family
(founders of Mexico's large private media empire Televisa) extended
its broadcasting operations across the border. At first this was in the
form of signals originating in Mexico that could be received easily
in border states. Once their television operations were underway in
the 1950s, however, the Azcárragas actually acquired affiliate stations in the United States – the first being a UHF station in San
Antonio. The Azcárragas' programming operation, Spanish International Network (SIN), was based in Mexico, but since the FCC
requires US broadcast stations to be owned by US citizens, the
family set up Spanish International Communication Corporation
(SICC) as something of a front operation, nominally owned by
American citizens (and eventually restructured following government investigations).

Over the next two decades, a number of other affiliates joined
the San Antonio station, forming the first US-based Spanish-
language programming network. In 1976, SICC/SIN fully interconnected all of its affiliate stations by satellite. While this
precedent-setting use of technology was not available to cable
systems at this date, the next move would be in that direction.
By 1979, SIN was paying cable operators at a rate of 10c per
subscriber with a Spanish family name for carriage of the signal.
Thus SIN was also one of the earliest satellite cable networks.
Others came soon thereafter. It also has continued to offer broadcast
service through its affiliate stations.

During the late 1980s and early 1990s, SIN and the SICC stations were sold to and resold (eventually being owned, among other interests, by Hallmark Cards). The somewhat reconfigured network was renamed Univisión (most of Univisión's US production operations are now based in Miami). Since that time, the network has become well known for its array of programs, ranging from news to *telenovelas* (serial dramas that run over a limited number of episodes). Especially popular have been the variety show *Sábado Gigante* and the talk show *Cristina*.

A second major Spanish-language network, Telemundo, entered the market in 1987. Some observers have noticed that Telemundo has been more popular than Univisión among US Latino/as not of Mexican descent. Telemundo, based in Florida, devotes a significant portion of its schedule to programming that addresses cultures of the eastern Caribbean nations (particularly Cuba and Puerto Rico). Meanwhile, Galavisión, an international satellite service launched by Televisa in 1979 and available in parts of the United States, was operating as a distant third in the competition for Spanish-speaking cable audiences.

The present-day success of these Spanish-language cable networks is testament to the fact that the United States now boasts one of the largest (and *the* wealthiest) Spanish-speaking populations in the world. Today, in an era when media users are thought of more as taste categories than as mass audiences, many different channels have been launched to address the television needs and desires of Spanish speakers as well as English speakers of Hispanic heritage. Moreover, quite a few of the more successful English-language cable/satellite networks – including HBO, CNN, and ESPN – have spun off Spanish-language sister networks.

passed. In April 1979, when Pinwheel was renamed Nickelodeon and uplinked as a commercial-free channel for children, it stood ready to change expectations for children's programming, just as other early satellite cable networks were changing expectations for their respective genres.

Local Programming

Public access programming also grew during the late 1970s. In addition to the fact that the FCC's 1972 Report and Order had mandated PEG channels for medium to large systems, which were increasing in number by this stage, the cable industry itself was using access channels as a lure when bidding for a community's franchise. As Laura R. Linder explains in a book chronicling the history of public access television:

> [T]he growth of public access television was assured simply because the growth of the cable industry itself had created a buyer's market. The competition between cable companies for local franchises was often intense. Opportunistic and forward-thinking communities were able to demand access facilities in exchange for awarding their franchises to certain cable operators. In an extreme case, Dallas-Fort Worth was able to get twenty-four access channels. (9)

Indeed it was during the late 1970s and early 1980s that some of the nation's most sophisticated and progressive public access facilities started up, including professionally staffed and well equipped production centers in communities ranging from Austin, Texas (with a population of some half a million at the time) to Sun Prairie, Wisconsin (with around 10,000 residents). It seems clear that, regardless of the FCC's mandate, the *success* of public access in any given community was the result of activist efforts during the franchising process and ongoing citizen participation after that.

These factors became even more significant following the 1979 *United States v. Midwest Video Corp.* Supreme Court decision, more commonly known as Midwest Video II. The Midwest Video II decision was complex, as it revisited the FCC's authority to regulate cable (the issue decided in the Commission's favor in the 1968 *United States v. Southwestern Cable* case, as discussed in Chapter 3). In a previous decision in 1972, *United States v. Midwest Video Corp.* (known as Midwest Video I), the Supreme Court had upheld the FCC's right to require cable systems with over 3,500 subscribers to provide access facilities and disseminate local programming. Nonetheless, when it revisited the issue seven years later, the Court reversed its decision, favoring the cable industry's position that the public access requirement violated both its First and Fifth Amendment rights.[4]

Figure 5.4 A production in process at KIDS-4, one of Sun Prairie, Wisconsin's two cable-access channels. Sun Prairie (population 25,000) boasts one of the nation's oldest and most sophisticated cable-access facilities, founded in 1979 [Sun Praire Cable Access]

We will never know whether or not more public access facilities would exist today had *Midwest Video II* not struck down the FCC's public access requirement. Without the necessary degree of citizen participation, an access facility can exist without actually being of service to its community. On the other hand, where a critical mass of residents has truly desired public access, facilities usually have become available through some combination of public resources and contributions from the local cable operator. Often, though by no means always, this has occurred as part of the franchising process. Since the late 1970s, most communities – large and small – have either negotiated a cable franchise for the first time or renewed an existing one. This is hardly surprising, since the arrival of satellites and later fiberoptic cable and other technologies has transformed the very notion of cable television service. Such technologies have made cable marketable where it would not have been if it had remained simply a community antenna service. They have also changed the expectations held by residents of those communities that do still rely on cable for community antenna service.

Further Consolidation of the US Cable Industry

In fact, the technological changes that began for cable in the 1970s were closely accompanied by changes to the structure of that industry, as ownership of more and more local cable systems – particularly those in larger metropolitan areas – came to be concentrated within cable MSOs. In the 1960s, MSOs such as Teleprompter and H&B were considered colossal, even though they controlled only a small fraction of the total number of local cable systems. But by the late 1970s and early 1980s, modern cable powerhouses such as Warner Amex, ATC, and Tele-Communications Inc. (TCI) were rapidly building their empires.

Teleprompter lost its extraordinary momentum in 1971 when CEO Irving Kahn was convicted of bribery during the Johnstown, Pennsylvania franchise renewal process. Even though Kahn devoted most of his prison term to earning a correspondence certificate in cable technology from Penn State University, this did not help him to regain either his or Teleprompter's former status in the cable industry. The company had already been in serious debt at the time of Kahn's conviction, and the damage to his reputation in the industry only hurt the company more. While Kahn was serving his term, Teleprompter was taken over by Jack Kent Cooke, whose own cable company had just merged with it. Cooke had Kahn's voting rights in the company withdrawn and made his associate William Bresnan, a one-time cable technician, president of Teleprompter's cable division. The two ran a greatly weakened Teleprompter during the late 1970s, just when satellite cable was becoming a reality and drawing subscribers in ever greater numbers. The company's stock continued to drop, in spite of efforts by Cooke and Bresnan to keep it out of debt. In 1981 they sold Teleprompter to Westinghouse's Group W Cable subdivision.

Meanwhile other MSOs were gaining a foothold in the industry. The early 1970s had been a difficult time financially for the entire cable industry, but most would enjoy a reversal of their fortunes as the decade went on. The later part of the decade saw the arrival of people with advanced degrees in cable-related fields, joining the mostly self-educated industry pioneers. In 1973 Robert Magness, founder and chairman of the regional MSO Telecommunications, Inc. (TCI), hired a new president, John Malone. Malone, still in his thirties, arrived with an M.S. in Industrial Management and a Ph.D. in Operations Research from prestigious universities. He also had been serving as president of Jerrold Electronics (at that point a division of General Instrument) and thus had experience in cable specifically. Malone's talents were sorely needed. As media journalist Thomas P. Southwick explains:

> Malone knew full well how bad the situation had become. TCI had been a major customer when Malone ran Jerrold, and the company had made full use of the financing program Malone had set up. When Malone moved to [the company's Denver Headquarters], TCI had $132 million in debt, and only $18 million a year in gross revenues. Its stock dropped as low as $.87 a share. Its market capitalization was down to $3 million. (143)

Under Malone, TCI would flourish. Among his strategies was to keep the company operating in the red, paying interest on loans, acquiring more assets, and writing off depreciation costs. To ward off any hostile takeovers, he reconfigured the company's stock: While "A" stock would continue to carry one vote per share, the premium "B" stock – available only to TCI management insiders – would carry ten votes per share. Cable has always been a heavily leveraged industry, but Malone turned debt management into an art form that not only saved TCI, but made it the largest cable MSO by the early 1980s. TCI was competing on a somewhat different playing field than other large MSOs, such as Warner Cable, ATC, and Westinghouse, in that it looked past the lure of big-city franchises and instead concentrated its efforts on the nation's numerous smaller markets. This was more affordable for the continually debt-ridden company, but it also wound up being a key to its long-term success.

Malone was also forward-looking and shrewd in his perceptions about the satellite-driven cable programming environment that was on the rise as of the late 1970s. He had watched HBO's satellite debut carefully and observed Ted Turner's aggressive programming maneuvers with fascination. Then, when untried programming ventures such as BET and the Discovery Channel needed venture capital, Malone was there. This was not exactly about altruism or even some vestige of Blue Sky era optimism, but rather a realization that from that point on, cable programming and system ownership would be intimately connected. Malone's biographer Mark Robichaux makes this clear:

> TCI systems reached 2 million viewers by 1981, officially making TCI the biggest cable operator in the country and far outpacing the nearest competitors; getting TCI to carry a new channel almost guaranteed its success. So when John Malone came calling, did the new channel really have any choice but to hand over a piece of the action? If they didn't let him become a silent partner, he might not allow TCI's cable systems to offer the new channel. Malone generally made the fledgling networks an offer they eagerly accepted, but also one they couldn't really refuse: for a small equity stake in the channel, usually well below a majority share, TCI would add the new service to its systems and give it a fast start in gaining critical mass. (61)

Of course, Malone was not the only cable magnate involved in the horizontal and vertical integration of the cable television industry at this stage. Others involved in gaining control of both distribution systems and content providers included Glenn Jones of Jones Intercable, Ralph Roberts of Comcast, John Rigas of Adelphia, Alan Gerry of Cablevision Industries, and several others. Cable was on its way to joining other media industries in the synergistic business practices of the late twentieth and early twenty-first centuries.

The International Scene

By the time of HBO's 1975 satellite debut, communication satellites had been in use for more than a decade – with functions ranging from long-distance telephony to surveillance to international transmission of television programming. The United States was hardly the only nation to be making strides with this technology. In fact, satellites proved extremely desirable in developing nations that were seeking to build their television infrastructures but lacked the technology and funding to do this terrestrially. In 1975, Indonesia became the first nation in the developing world to launch a communications satellite. The satellite, *Palapa*, initially was meant to facilitate business communications between far-flung industrial operations and the national capital, Jakarta. However, Palapa was quickly enlisted by the national broadcaster, TVRI (Televisi Republik Indonesia), as a means of distributing its signal to the various islands which comprise the nation. Although launching a satellite was extremely expensive at this early stage, it was considered far more economical than constructing the extensive microwave relay network that would have been the alternative (Thomas: 130). The Indonesia case, and those of other nations to follow, makes it clear that relative economic prosperity need not be the only factor in the adoption of new communication technologies.

For this reason, during the 1980s and 1990s several developing nations would lead the way in forming transnational networks that would serve regional or linguistic communities. Indonesia and some of its South Asian neighbors would spawn such enterprises, as would a handful of larger Latin American nations. Mexico's large private broadcaster Televisa is one example. As John Sinclair explains:

> In view of the fact that the name Televisa was conceived as an abbreviation for Televisión vía Satélite, it would come as no surprise that much of the technological development which was undertaken in these years had to do with Televisa's ambitions to extend its distribution to selected audiences on an international scale. Satellite distribution has thus augmented the role of Protele, Televisa's sales division, which, as well as selling Televisa programmes to the few independent channels remaining in Mexico, exists to sell programme rights as well and physically distribute programmes on an international basis through its foreign offices. In 1976, Televisa commenced a venture it called Univisión, a weekly feed of its domestic programming to the US border for subsequent distribution to its stations in the US via satellite. The following year, Televisa established an office in Madrid, and began occasional satellite transmissions to Spain. (40)

Televisa's efforts in developing the transnational PanAmSat satellite network, as well as the company's subsequent ventures in a variety of other pay-television delivery systems (cable, DTH satellite, MMDS), helped Mexico to become the second largest exporter of television programs (in terms of both sales and distribution) in the late twentieth and early twenty-first centuries.

If necessity was "the mother of invention" (or of satellite entrepreneurship) in the developing world, however, developed nations with longstanding terrestrial television infrastructures (particularly those involving public service broadcasters) were somewhat more cautious in their approaches to new multichannel delivery technologies. Several European countries, accustomed to only one or a few publicly funded broadcast channels, enjoyed their first wave of expansion in the 1980s merely by allowing new channels into the market (either additional public channels or, more commonly, newly permitted privately owned competitors). In most situations, DTH satellite, cable, and other pay-television options would not see much success until well into the 1990s, once the concept of expanded television service had been planted in consumers' minds. And even though Canada always has been one of the most heavily "cabled" nations for CATV service and had been able to offer its Anik satellite for HBO's 1973 Anaheim demonstration, the country faced some regulatory difficulties when getting its own satellite-served cable industry off the ground. With the exception of some trans-border piracy of US channels, pay-cable services did not became available to Canadians until the mid-1980s.

In all, the late 1970s would be the last years in which the world would have doubts about the future of cable and satellite television technologies. People in general were still uncertain as to what these technologies meant for the future of existing television infrastructures – whether the commercially supported broadcast networks of the United States, the public service broadcasters in Western Europe and elsewhere in the developed world, or the fledgling television operations of many third-world countries. But it was known universally that changes were in the making. The next chapter will look at how these changes began to manifest themselves, as well as how policy-makers began to contend with them.

NOTES

1 The complete text of the Act and its amendments can be found at the US Copyright Office website: http://www.copyright.gov/title17/.
2 Dolan then acquired the former Sterling systems and founded Cablevision Systems Corp., which would quickly grow into one of the nation's largest MSOs.
3 These would be joined on satellite by New York's WPIX and WNYW, Boston's WSBK, and Los Angeles's KTLA during the 1980s. Each of these stations was already being carried regionally by microwave.
4 In other words, the First Amendment was seen as protecting cable operator's programming decisions as a form of free speech. The Fifth Amendment was seen as protecting their freedom from seizure of property without due process of law.

FURTHER READING

Linder, Laura R. *Public Access Television: America's Electronic Soapbox*. Westport, CT: Praeger, 1999.
Mullen, Megan. *The Rise of Cable Programming in the United States: Revolution or Evolution?* Austin, TX: University of Texas, 2003.

Chapter 6

The Satellite Years:
1980–92

If the late 1970s saw the shift of cable from a retransmission medium to a programming medium, the decade that followed set in place the notion that cable could and would become a definitive competitor for broadcast television. Not only did the handful of satellite networks launched after 1975 solidify their marketing identities in the 1980s, dozens of new networks got started during that decade as well – representing both successes and lessons learned through failure. If cable had been an obscure rural retransmission medium until the late 1970s, this certainly was no longer the case during the 1980s. By the early 1990s, virtually every community in the United States had been wired. This, of course, included the larger broadcast markets, where retransmission of broadcast signals alone was hardly sufficient reason to subscribe to cable service. Success in these markets indicated that cable would be a formidable presence for the foreseeable future.

The period from 1980 to 1992 also saw passage of the first *laws* to affect cable directly: the 1984 Cable Act and the 1992 Cable Act. Although cable remained a subject of great uncertainty for legislators, the industry's growing power and influence were palpable. It was clear that the industry was in need of more oversight. The two successive bodies of legislation, unfortunately, would take both the cable industry and the cable-subscribing public on a roller-coaster ride that was anything but reassuring to either. Congress wound up, first, giving nearly free rein to the cable industry and causing consumer costs to increase dramatically and then, later, placing restraints on the cable industry that hindered their business but did little to help consumers. Government policy-makers simply could not keep pace with the speed of technological innovation and the demands that innovation placed on cable operators. They also did not perceive how much the young cable industry differed from more established industries.

As of 1980 cable was grappling with its new role as a provider of multichannel television service. Both operators and programmers were trying to figure out how to present the television-viewing public with something that differed from what was on broadcast television, but this had to be done with very limited budgets – especially when compared with the large budgets of the established broadcast networks. Conveniently enough, the American audience responded favorably to the syndicated reruns most fledgling cable networks had to rely upon to fill out their schedules. During the 1980s, cable began to become known as the television home for "classic TV."

Then in the early 1990s, the young industry had to begin another shift in direction. Even while several networks were making names for themselves as purveyors of quality original programming, the cable industry suddenly realized that it could no longer remain in the business of providing television programming *alone*. Cable was widely perceived as the medium to provide a futuristic array of interactive services – even though the specific nature of that interactivity was not entirely clear. Moreover, competition was just beginning to emerge from telephone companies, which, newly free to compete in certain cable-related arenas, also sought to provide the highly desired new services.

Also, by 1992, the cable industry had other competition. The period discussed in this chapter can be designated as cable's "satellite years" not only because of the extent to which the advent of satellites had affected cable's programming role, but also because these years saw the rise of other satellite-dependent multichannel television technologies such as satellite master antenna television (SMATV), multichannel multipoint distribution service (MMDS, or "wireless cable"), and, most prominently, direct-to-home satellite (DTH). The cable industry definitely held its own, however. In fact, its dramatic growth was very much in the public spotlight at this stage.

Big Players Getting Bigger

During the 1980s, MSOs were growing wealthier and more powerful almost by the minute. Among them was the behemoth TCI. TCI's strategy of buying up smaller cable systems while its rivals went after the larger markets paid off as early as the middle of the decade, by which point Malone and his associates were prepared to begin acquiring larger cable systems. Among other things, they went after systems that were failing because their owners had promised too many bells and whistles during the franchising process. One of these was in Pittsburgh, a system acquired by the Warner Amex partnership a few years earlier when they were promoting the unsuccessful interactive QUBE system as the wave of the future. As Mark Robichaux explains:

> Warner Amex let the system go for $93.4 million in cash in 1984. Malone then quickly introduced the city to TCI's unadorned, and on-the-cheap service. Instead of the elaborate cable system of the future that Warner Amex had promised, Malone renegotiated the franchise to offer a plain-vanilla system that would let it cut costs, yet continue to charge essentially the same rates as Warner Amex had charged . . . TCI promised a system with 44 channels instead of 63; it ripped out the QUBE system and the expensive equipment supporting it and sold the interactive boxes back to Warner for $30 each. Malone cut payrolls by nearly half, closed the elaborate studios promised to local officials, and moved the extravagant downtown headquarters to a tire warehouse. Pittsburgh was a steal for TCI because within two years, TCI had lowered debt and improved finances enough to begin paying back banks using the system's own improved cash flow. (76)

The central Pittsburgh system was then easily connected to smaller TCI systems in the suburbs and the outlying region. This, of course, was precedent-setting. Other MSOs quickly perceived the economic benefits of such centralization and jumped on the same bandwagon.

Regional consolidation centralized such formerly local functions as billing, marketing, and headend operations at an urban hub with streamlined staffing. This carried the additional advantage of a larger and more skilled applicant pool for the various jobs. As cable ownership grew increasingly concentrated over the next decade, the regional service areas would only get bigger. ATC (in the years immediately before it became part of the Time Warner empire) was the first to pursue the urban hub aggressively, under the direction of CEO Trygve Myhren, who explained:

> I actually enunciated something called the clustering strategy. This was
> a way to take our business forward more powerfully than the scattered
> cable system model. It was really to buy contiguous properties and to
> try to fill out an entire media market. A media market can be described as
> something that might be the outline of the newspaper's delivery or the
> Grade B context [*sic*] of the television stations. Usually also it was fairly
> contiguous with radio station coverage. But basically, the footprint of
> the traditional media – we saw that as a cluster. (Myhren oral history)

Other MSOs quickly followed suit.

Meanwhile, competition for cable franchises had reached fever pitch
by the early 1980s. Negotiations between city councils and would-be
franchise holders had grown corrupt to an extent that made Irving
Kahn's actions in Johnstown, Pennsylvania, for which he served a
prison sentence a decade earlier, seem rather benign by comparison.
The publicity surrounding new cable-related services and technologies had prompted many cities to insist on lavish, state-of-the art
systems and well-equipped public access facilities in their franchise
agreements. But even beyond exorbitant (but still legitimate) cable-
related demands, some were asking cable companies to pay for public
works projects such as tree planting in exchange for franchises. Some
cable companies even engaged in a corrupt practice known as "rent-
a-citizen," in which city council members and other influential local
people would be flown to popular vacation destinations such as Las
Vegas for informational sessions and technology demonstrations.

Of course, not all cable operators were actively engaged in border-
line bribery – but any wishing to succeed in the satellite era realized
they needed to be more aggressive than in the past. Malone, while
openly rejecting the corruption evident among his peers, is said to
have been fairly belligerent in TCI's franchise negotiations, nonetheless. He proved to be a thorn in the sides of nearly all local city
councils with which TCI did business during the franchise renewal
process. Malone and his associates waged nasty legal and personal
campaigns against city officials, and even though TCI seldom won
these battles outright, they drained the energies and resources of all
but the most stubborn local constituencies, causing them to concede
defeat (see Robichaux: 61–72). These practices eventually caught the
attention of federal legislators, adding to other concerns they had
about the cable industry.

The 1984 Cable Act

Formally known as the Cable Communications Policy Act, the 1984 Cable Act was the first comprehensive cable legislation ever to be passed – in spite of fairly regular efforts to do so, dating back to the early years of CATV. As discussed in previous chapters, in the absence of legislation, cable had fallen awkwardly under the FCC's regulatory authority as established in the 1934 Communications Act – long before the first community antennas even existed. While the Supreme Court's 1968 decision in *Southwestern Video v. FCC* had authorized the FCC to regulate cable, much more was needed.

The 1984 Act (and the various bills, court cases, and other policy documents that preceded it) was aimed, first and foremost, at reaching a compromise between the interests of the cable industry and those of the municipalities in which cable systems wished to do business. With the rise of satellite cable and the services it promised, it had become apparent that the financial stakes were high for both sides. Clearly the franchising process already had become riddled with corruption. By 1984, a few concerned states had passed their own legislation in efforts to rein in the franchising processes, and so Congress eventually took on the issue at national level. The key lobbying organizations involved in framing the legislation were the National League of Cities and the NCTA (as well as several smaller cable trade associations). Citizens groups also weighed in, though with relatively little influence.

At this stage of political history, there was every indication that any cable-related government policy would strongly favor the growth of the industry. The deregulation-minded Ronald Reagan had been elected President in 1980 and had subsequently appointed big business advocate Mark Fowler to head the FCC. Republicans also held a majority of Senate seats. Furthermore, the Act was very much the personal project of Colorado Congressman Tim Wirth, chair of the House Subcommittee on Telecommunications. Although a Democrat, Wirth's liberal-centrist position complemented the dispositions of the Reagan-era Republicans when it came to deregulation of the cable industry. A number of his constituents were involved in cable and many cable companies were headquartered in his state. For a number of years, Wirth had espoused the position that less-regulated

Figure 6.1   Tim Wirth [The Cable Center]

competition would open the telecommunications industries to innovation and ultimately benefit consumers. He was a lingering believer in Blue Sky – seeing those ideals coming to fruition, by this stage, in the form of specialized cable services such as C-SPAN, BET, Nickelodeon, and CNN.

Most provisions of the 1984 Cable Act thus reflected a much larger trend toward deregulation of US industries generally during the 1980s. In the Act, cities won something of a pyrrhic victory. They gained the right to require franchises, even though franchises were largely *pro forma* by this stage, anyway. They gained the right to charge franchise fees, even though these were capped at 5 percent – less than many already were requiring at the time. They also were given leased access channels, though these would be used far less than anticipated. And finally, cities gained some control, at least in theory if not in practice, over programming "decency" standards – an issue on the public's mind since some cable networks already were showing some mildly "adult" material. But the Act ultimately did *much more* for the increasingly powerful cable industry. It freed operators from most types of control at the municipal level and protected their existing operations from prosecution for all but the most egregious forms of negligence or business misconduct. And it gave them the go-ahead to increase rates – which they did, extensively, over the coming years.

Of course, this deregulation benefited an industry that was increasingly concentrated and thus enjoying the economies of scale that accrue to those who own many systems. The larger MSOs were able to raise subscriber rates even while consolidating operations and negotiating huge bulk discounts from cable networks. Ultimately consumers would be hurt the most by the 1984 Act – something that federal legislators would start trying to address almost immediately following its passage and culminating in the 1992 Cable Act, discussed below.

Cable Networks

Along with easing the franchising process for cable operators, the 1984 Act helped the cable industry expand its programming operations considerably. The 1980s and early 1990s saw the number of cable networks grow almost exponentially. In 1980, there were around twelve networks, depending on how a network was defined.[1] Twelve years later, there were approximately five times this many. Where earlier cable networks had largely come about through a combination of established programming operations and entrepreneurial risk-taking, this second wave was much more calculated. Aspiring programmers had to determine which programming niches were both underserved and likely to draw enough interest to pay off (in the form of subscriber fees and advertising revenues for basic networks and in the form of subscriptions for premium networks). They then had to convince operators to carry them – a difficult proposition, since the growth in the number of cable network startups far outpaced the increase in individual cable systems' channel capacity (which averaged around 30–40).

The most notorious programming miscalculations of the 1980s were embodied in the so-called "culture" networks. Once it was acknowledged that satellite cable was a viable enterprise, some entities with deep pockets set out to fulfill one of the lingering Blue Sky dreams: channels that would deliver live cultural performances (stage plays, ballets, concerts, etc.) and art films directly to home audiences.

Each of the three existing broadcast television networks made a foray into cable's "cultural" programming arena; they obviously saw some potential in satellite cable and were hoping to carve out some territory for themselves. ABC had ABC-ARTS (Alpha Repertory Television), CBC had CBS-Cable, and NBC had the Entertainment Channel. In addition there was Bravo, the founding network of Rainbow Media Group, the programming division of the MSO Cablevision Corporation.

All of these networks floundered in the beginning, though; CBS-Cable simply failed outright. It seemed that the high culture-oriented audience – traditionally served by PBS – was not interested in subscribing to cable for these networks alone and had little interest in the other networks available at the time (see Waterman for more discussion of this). It cannot be said that the cultural networks were a complete failure, since most of them survived in some form or another. But it did become clear that survival on cable television in the United States called for some compromises – namely finding a middle ground between lofty aspirations of delivering high culture to people's living rooms and the realities of both industry economics and the broad television audience's established tastes and expectations.

Parts of ABC-ARTS and the Entertainment Channel were reconfigured to form Arts & Entertainment (A&E), still a popular channel in today's cable line-ups. However, A&E hardly developed as the source of traditionally defined high culture that its founders had anticipated. Rather than putting large amounts of money into covering prestigious live culture events, A&E chose to create inexpensive "signature" documentary programs, notably the popular and long-running *Biography*, and subsidize production of these with inexpensive off-broadcast crime dramas such as *Quincy* and *Columbo*. Bravo had an easier time maintaining the niche its founders originally selected. From the start, Bravo opted for the foreign and independent films seldom seen elsewhere on US television, or even in most movie theaters. As early as the late 1950s moviegoers in urban areas and college communities had developed a taste for non-Hollywood films, but there were relatively few places to view them. Bravo not only changed that situation, but also enhanced the experience of watching them on cable by filling in the odd time slots between feature-length films with short films (also seldom seen in other venues), in-studio interviews, and film-related documentaries.

Other innovative and popular cable networks with their origins in the 1980s include Discovery and The Learning Channel. Discovery launched in 1985, as the brainchild of John Hendricks, a consultant in higher education. As he explained in a 1993 interview, "I realized that nothing substantive was being shown on network television . . . The problem is that when you're driven by ratings and advertising, you end up losing the more discriminating viewer" (Merina). Hendricks not only perceived cable as a home for documentaries and other sorts of educational programming to serve this underrepresented audience; he was aware of a huge supply of this sort of programming at a low cost from other nations' public service broadcasting companies (Canada's CBC, the UK's BBC, and others). Several MSOs, seeing Discovery as a good public relations tool (especially among educated, upscale viewers), lent support in the form of both financing and carriage. Group W, which had remained in the cable programming business after selling off its cable systems, also was a major investor in Discovery.

Meanwhile, Discovery's now-sister network, The Learning Channel (TLC), had emerged from some different origins. The Appalachian Community Service Network (ASCN) had formed in 1972 for the purpose of using NASA's experimental communications satellites to deliver educational programs to communities in isolated and economically disadvantaged parts of the United States. In 1978, by which point the satellites were no longer just experimental, ASCN had secured permanent transponder space on RCA's SATCOM III satellite. It was then able to transmit its adult-targeted documentary programs nationwide. In 1986, ASCN (with the help of other investors) spun off TLC as a for-profit subsidiary and three years later TLC was offering a 24-hour program schedule. In 1991, TLC was acquired by Hendricks and Discovery Channel. The two-network company, Discovery Networks, quickly acquired or developed a host of other cable networks (analog and eventually digital), including the Travel Channel, Animal Planet, Discovery Health Channel, and Discovery Home and Leisure. By the late 1990s, "Discovery" would be a brand encompassing not only cable networks (both in the United States and internationally), but also retail stores, clothing, and home audio and video products.

Quite a few other successful cable programming operations that began during the 1980s and early 1990s were launched by individuals and companies already established in television programming – many in cable programming specifically. They had some financial resources as well as reputations from which to draw. Moreover, they were able to adjust subscriber fees for existing networks in ways that made it financially desirable for operators to find room for the start-ups. For instance, drawing on HBO's established reputation with the television audience and its promising revenue potential, Time Inc. was able to launch Cinemax, an all-movie network, in 1980. Cinemax was intended to complement HBO's programming mix and thus replace Showtime or The Movie Channel as subscribers' second pay-cable channel, and for a while operators were offering discounts for subscribing to both Time Inc. networks (a practice that was later deemed anticompetitive).

Lifetime ("Television for Women") had a more complicated developmental trajectory. Similar to A&E, Lifetime was the result of two networks, each of which had proved to be a little too specialized. One was Daytime, a joint venture of the Hearst Corp. newspaper chain and the ABC broadcast network, that targeted women during weekdays only (1:00–9:00 PM). The other was Viacom International's Cable Health Network (CHN), with a selection of both general interest and specialist-oriented medical programs. The merged Lifetime network, following its debut in early 1984, tried to retain as many elements of Daytime and CHN as possible. It was not difficult to keep the Daytime elements in place; not only were there long-established precedents for targeting women during weekday hours, it was also possible to find corporate sponsors to pay most of the production costs for such shows as *What Every Baby Knows* (produced by Procter & Gamble), a parenting advice program featuring well-known pediatrician T. Berry Brazleton. But the medical programs, particularly those intended for medical professionals, were draining resources. In spite of praise from the medical community, Lifetime eventually shifted its Sunday medical lineup to entertainment programming – emphasizing reruns of primetime series (such as *L.A. Law*) with widespread appeal among both men and women.

Clearly cable lacked the channel capacity to accommodate true niche audiences at this stage. This was not a wholesale defeat of the Blue Sky ideals, however. Rather it was an indication that a complete or immediate overturning of the existing economic structure that drove US television – and the audience expectations that had been cultivated within that system – was not feasible. By the mid-1990s, there would be evidence that the rise of satellite cable had indeed brought with it a shift toward more specialized television networks – as will be discussed in Chapter 7. In the meantime, cable entrepreneurs were enjoying some notable successes with more traditional television genres.

For one, self-proclaimed populist Ted Turner was active in starting more new networks during the 1980s and early 1990s. Pleased with the success of his TBS superstation, he went on to launch Cable News Network (CNN) in 1980, taking on a television genre in decline. The broadcast networks, whose news bureaus produced most national and international news, saw it as a drain on revenues. And not as many people – especially young people – seemed to be watching the news. Turner chose to gamble on cable's potential both to draw people back to the news and to change viewing behaviors. He clearly was taking a risk. In its early years, CNN was sometimes referred to as the "Chicken Noodle Network." The production values in its news programs were notoriously low – among the lowest among all cable networks owing to the need to offer a schedule comprised entirely of new and original programming.

Turner found a niche for CNN in picking up the international news coverage the broadcast networks had begun to drop. Journalist and Turner biographer Ken Auletta explains: "By the mideighties, when each of the [broadcast] networks was taken over by giant corporations that focused more on cost cutting and driving up the stock price, the networks began to close overseas bureaus. With the end of the cold war, interest in international news coverage waned. There was no Al Jazeera, no worldwide BBC as we know it today" (43–4). As the broadcast networks downsized, Turner built up CNN's international infrastructure. CNN thus was well positioned to satisfy the public's thirst for news of the 1990 Gulf War. According to Auletta, "CNN made a financial commitment to cover any potential war from ground zero, Baghdad . . . CNN invest[ed] in a suitcase version of a satellite phone; this expenditure, of fifty-two thousand dollars, would be added to the extra millions that CNN was committed to spend to cover what might be the first live war" (53).

Turner also had launched CNN Headline News in 1982 as a provider of regularly updated half-hour news summaries. And, never neglectful of the public's demand for pure entertainment, he added

Turner Network Television (TNT) to his programming empire in 1988. Cartoon Network would launch in 1992 and Turner Classic Movies (TCM) in 1994. These would later be joined by the "classic" cartoon network Boomerang, the regional network Turner South, and several international versions of the company's various networks (including CNN International in 1992). One of the most frequently cited factors in Turner's cable programming success has been his personal love of the content, particularly the older movies and sports events, and his personal commitment to making it available. Another factor has been his direct acquisition of popular programming sources – from professional sports teams in Atlanta in the 1970s to his purchase of the MGM film library in 1985 (the one portion of the movie studio he was able to retain following a disastrous acquisition and short-lived ownership of the entire company).

Of course, the new cable networks lacked broadcast television's programming budgets. But, as discussed above, it also was known that television reruns have enduring popularity with the television audience. And so, during the 1980s, some interesting and resourceful strategies emerged to reinvigorate and recontextualize secondhand syndicated fare and to save money on producing or acquiring original programs. In this vein, Nickelodeon launched a nighttime subservice called Nick at Nite, designating it a home for "classic" television. Nick at Nite initiated the practice of surrounding the old reruns that filled its schedule with clever commercial bumpers and promotional spots that made affectionate jokes about those programs' clichés and outdated stylistic features. Nick at Nite also introduced marathons, in which episodes of entire series are run back-to-back over entire days, and theme days, featuring episodes of different programs with common themes (e.g., holiday episodes). Still popular, these strategies are meant to flatter couch potatoes and devotees of specific programs alike.

Beginning in the early 1990s, some networks actually would dissect reruns (particularly those in non-narrative genres such as comedy and variety shows) in order to combine segments from different series within a particular theme, such as political election humor. This is still a common practice for Comedy Central and E! Entertainment Television, to provide a few examples. Another strategy Comedy Central in particular used during its lean start-up years involved the use of voice-overs and image-overs to make jokes about whatever inexpensive

Figure 6.2 Promo from Nick at Nite's September 1992 "Marython," in which all episodes of the long-running sitcom were shown back to back [author photo]

programming they were using. One example of this was its *Mystery Science Theater 3000*, which recycled old, low-budget movies using the premise of a man marooned in a spaceship with no companions other than a trio of "robot friends" who would watch "cheesy" movies with him. This comically implausible combination of viewers would be superimposed in silhouette across the bottom of the screen, with their mocking commentary audible over the original soundtrack (see Mullen).

One cost-efficient way of acquiring *new* programming on a tight budget was for a cable network to enter into a production partnership with a company seeking more advertising time than the traditional 30-second spot. In an arrangement reminiscent of the single-sponsor production style the broadcast television networks inherited from radio and continued to use through the mid-1950s, cable networks would offer half of the advertising time during a program to a company paying all or part of that program's production costs. Of course, the

Figure 6.3 On Comedy Central's *Mystery Science Theater 3000*, the hapless Joel watches movies with his "robot friends" [author photo]

program's content would be tailored to coordinate well with the sponsor's messages. Lifetime's program, *What Every Baby Knows*, mentioned above, is one example. A much more prominent example of commercial sponsors supplying a cable network with programming was MTV's groundbreaking video "clip" strategy, with popular songs accompanied by live concert footage or short dramatic performances (a number of which were the work of well-known Hollywood movie directors). Even though the clips were remarkably popular with teenagers and young adults (groups traditionally difficult for television networks to reach), they were literally nothing more than extended commercials for singles and albums (see Box 6.2). Black Entertainment Television (BET) also built its reputation, in part, by using music video clips.

The wave of home shopping channels that emerged during the late 1980s represented yet another source of inexpensive – and extremely

Box 6.1   Robert L. Johnson and Black Entertainment Television

One extraordinary figure in cable television history was Robert L. Johnson, founder of Black Entertainment Television. Born in 1946 in Hickory, Mississippi, Johnson was in his early thirties when he got the idea for Black Entertainment Television (BET), the first television network specifically for African Americans and one of the very earliest satellite cable networks. Clearly he was at the forefront in his perception of cable's potential to target specific audience segments.

Johnson had begun his career in cable in 1976, as the NCTA's Vice President for Government Relations. With a bachelor's degree in political science from the University of Illinois and a master's degree in international relations from Princeton University, Johnson hadn't exactly planned a career in the cable industry. The NCTA

Figure 6.4   Robert L. Johnson [RLJ]

position had been suggested to him by a Washington, DC neighbor who happened to be the secretary to the organization's president, Robert Schmidt. Interestingly, one of his responsibilities in this capacity had been lobbying to end the heavy restrictions on pay-cable.

Johnson benefited from connections made in the NCTA – people like John Malone, Robert Rosencrans, and Kay Koplowitz. The biggest initial investment in BET came from Malone. At first Johnson had been nervous about approaching the TCI mogul for assistance, especially since he knew Malone's political views differed from his own. But eventually he did so anyway, explaining in an interview years later:

> I went in and I started talking to him about this concept, and John, interestingly enough, when we talked about it had the cable system in Memphis, Tennessee, shared with ATC, I think, at the time, and they were looking for ways to get programming in to sort of augment their franchise proposal and to be able to say we're going to bring more programming in. He said, "Hey, you know this idea of putting programming on the satellite would help me solve my distant signal problem. So, yeah, if you can get programming . . . can you get programming?" I said, "Yeah, I think I can get programming." He said, "Well, if you can get programming, I think I'd be interested in helping you out, being your partner in the deal." I said, "Okay, when I come back to you with something I'll lay it out for you." (Johnson oral history)

Malone ended up investing $500,000 in BET. Johnson had secured some additional funding, though not much. The network began with a two-hour programming day, consisting of inexpensive old movies and gospel music clips. Facilities were rented in northern Virginia.

At first, BET was one of several networks sharing the Madison Square Garden Network's transponder – due mostly to the assistance of Rosencrans and Koplowitz. Other networks on the transponder included MSG itself, a popular sports channel; Calliope for children; and C-SPAN. Then in 1980, MSG and Calliope were combined and reconfigured to form USA, while C-SPAN and BET went onto their own transponders. BET began 24-hour service in 1984.

Like most start-up cable networks of its era, BET began with inexpensive broadcast reruns such as the sitcom *Benson*, old movies, and inexpensive original talk and variety shows. It also benefited from the popularity of the new music video format, particularly since MTV was giving virtually no attention to the increasingly popular rap music genre at this stage (emphasizing instead rock and new wave music). In fact, BET eventually was able to spin off three digital networks specializing in historically African American music genres: BET Gospel, BET Hip-Hop, and BET Jazz: The Jazz Channel.

The BET networks were acquired by the Viacom Inc. media conglomerate in 2000. As of 2005, BET had 80 million cable and DBS subscribers in the United States, Canada, and the Caribbean. In addition to media ventures, the BET brand also extends to restaurants and consumer products. Johnson continues to serve as Chairman of the Board at BET and in 2003 became owner of the Charlotte Bobcats NBA franchise.

---

successful – cable programming. Even though the QUBE version of interactive cable tested during the late 1970s and early 1980s had not worked out, home shopping was one version of interactivity that was gaining a foothold on cable during the 1980s. As cable services go, home shopping actually has not been all that sophisticated technologically. Sales have generally been conducted over the phone, with credit cards. The concept is reminiscent of the direct-response advertising trend begun in the late 1960s and early 1970s by the Popeil family and others – with products such as the Ginsu knife, the Buttoneer, and the Salad Shooter. Most of cable's home shopping channels have relied on low-budget studio set-ups, with live hosts expounding the virtues of gadgets, collectibles, inexpensive jewelry, and other products. They have given cable operators fairly generous profit-sharing incentives to ensure continued carriage in spite of shelf space limitations.

The home shopping concept has not been exclusive to cable television. The first home shopping service, the Home Shopping Channel, actually got started on a Clearwater, Florida radio station in 1977. Station owner Lowell "Bud" Paxson (who later founded the PAX broadcast network) agreed to accept 112 electric can openers from one of his advertisers in lieu of a cash payment. He was able to sell the can openers quickly over the air, and a few years later moved the

Box 6.2   MTV: Building the Brand

MTV-Music Television is one of cable television's earliest and most enduring success stories. In its quarter-century of operations, the popular basic cable network has managed not only to stay current with trends in youth culture around the world, but also to play a major role in shaping those trends.

When launched by Warner-Amex in 1981, a major goal for MTV was to appeal to teenagers and young adults, a demographic segment not well served by broadcast television. Yet, like other early satellite cable networks, MTV faced the challenge of filling its program schedule on a fairly low budget. One of MTV's founders, John Lack, had been Warner Cable Corporation's (WCC) executive vice president of programming and marketing in 1979, when the American Express corporation purchased 50 percent of the company and set out to boost existing programming entities (notably Warner Star Channel, which would become the Movie Channel, and the portions of the experimental interactive QUBE systems that would become Nickelodeon) and develop new ones.

In 1981 Lack, who already had a strong background and interest in popular music, hired a young market research expert, Robert Pittman, away from CBS to head up Warner-Amex's pay-TV division. Lack and Pittman together have received the most credit for launching MTV. The music video format was not entirely the innovation of MTV's founder's, though. Arguably the concept dates back to the early rock 'n' roll movies that featured such performers as Elvis Presley and the Beatles. However, during the mid- to late 1970s, some performers (notably Todd Rundgren) and recording companies had begun experimenting with short promotional music video "clips." It just wasn't entirely clear where or how these would be used. The music recording industry had been in a slump following the decline of disco, and so when Lack and Pittman presented the idea for an all-music cable channel featuring those clips, they had a fairly receptive audience (see Denisoff for more details on the founding and early years of MTV).

And as they say, "the rest is history." Not only would MTV capture a lucrative audience segment among cable subscribers, it would deliver that group to advertisers of consumer goods such as soda pop and chewing gum. It would be, in other words, an advertising vehicle for recorded music – and a whole lot more! MTV also built on the radio disc-jockey role by employing video jocks, or "veejays," including the infamous "Downtown" Julie Brown.

MTV's music video-clip format proved to be a founding genre for cable television both directly, as some networks began to dedicate portions of their schedules to music clips, and indirectly, as other networks adopted the short and flexible format to genres such as sketch comedy (e.g., Comedy Central) and talk shows (e.g., E! Entertainment Television). However, no sooner had MTV itself cultivated a taste for the music video clips than it began to expand its program repertoire. By the late 1980s, it had added thematic programming blocks, including *Club MTV* (dance music) in 1985 and *Yo! MTV Raps* in 1988. It had been running *MTV News* since 1981 and added the game show *Remote Control* in 1987.

Programs such as these marked the beginning of MTV's metamorphosis from an all-music channel into a home for youth culture generally (spinning off its sister network VH1 in 1985 to target a somewhat older demographic is also evidence of this). This is apparent in the success during the 1990s and early 2000s of such non music-specific programs as *The Real World* (debuted 1992 – a very early example of the reality television genre) the animated *Beavis and Butt-head* (debuted 1993), and *Pimp My Ride* (debuted 2004). Andrew Goodwin argues that MTV's scheduling shift to more demographically-predictable "dayparts" was an acknowledgment that in spite of its innovative new programming thrust, it still had to exist within the established world of commercial television (142–9).

Whatever its current programming and marketing strategies might be, one thing has been true of MTV's success: its ability to promote "name brand" or "signature" programming. This has been reinforced through its clever logos, introduced at the network's inception, and animated ID spots (see Rabinovitz). It has also been apparent through publicity campaigns such as "Spring Break" and "Rock the Vote."

MTV's inextricable connection with youth culture now extends far beyond the borders of the nation in which it was founded. Its brand has been franchised in spin-off networks around the world (in addition to more narrowly targeted sub-networks such as MTV Español in the United States). MTV Networks (now a division of the Viacom media conglomerate) has even announced plans for a Middle Eastern music channel, MTV Arabiya, to be launched in

2007. This is remarkable, given the various Islam-based restrictions on American-style pop music. And even where national policy prohibits the operation of a US-based network, MTV-lookalike networks have been created to infuse MTV-style programming with indigenous content. In Canada, for example, where the CRTC enforces its Canadian Content quotas, the strikingly MTV-like MuchMusic network does a booming business. Clearly, there is no escaping MTV's powerful influence.

Figure 6.5 QUBE was an experimental interactive cable channel that ran on Warner Amex's Columbus, Ohio cable system during the late 1970s and early 1980s. *Qubesumer* was one of the short-lived programs offered on the channel [The Cable Center]

home shopping concept to a public access channel in Tampa. In 1986, the renamed Home Shopping Network became available nationwide via satellite – distributed through *both* local cable systems and independent broadcast stations. It was quickly joined by competitors QVC and Valuevision. By 1990, "home shopping" was a buzzword in the cable industry and beyond – with parties as disparate as Ted Turner, Playboy Enterprises, and Macy's department stores all trying to make home shopping a feature of their existing lines of business. The intensity of the trend did not last, though. There are far fewer home shopping channels available on cable today than there were in 1990. No doubt this is evidence of the greater variety and convenience available to at-home shoppers from the Internet – whether through dedicated shopping sites such as amazon.com or directly from manufacturers and distributors via their company websites.

Meanwhile, cable networks that had launched during the 1970s were now able to offer other, often more costly, sorts of original programming in their second decade. Nickelodeon was one of these, having developed a schedule of edgy new children's programs that generally were liked by both kids and their parents. Programs such as *Double Dare*, *Clarissa Explains It All*, and *Rugrats* found a niche in fostering a kind of benign irreverence toward the adult world. Founder Geraldine Laybourne reflects that "It was clear to us in the mid-80s these kids were being so pressured to grow up fast that they were missing a childhood. If Nickelodeon could become a place where they could just be themselves and be liked for who they were, and have it be playful and silly and not so earnest, that we would be doing a great service for kids" (Laybourne oral history).

HBO began producing highly acclaimed original movies during its second decade, including *The Terry Fox Story* (1983), *Common Threads: Stories from the Quilt* (1989), and *The Josephine Baker Story* (1991). And as its fortunes grew, HBO also increasingly contributed to the costs of producing Hollywood movies – a wise move, as this gave the cable network increasingly valuable first television distribution rights. Its first "pre-buy," in 1980, was Universal's *On Golden Pond* (released in theaters in 1981). In 1982, HBO joined with Columbia Pictures and CBS to form TriStar Pictures (now part of Sony Pictures) as a vehicle for upfront agreements that would secure television exhibition rights. From this point, the number of major Hollywood movies that made their way to HBO and other pay-cable

Figure 6.6   Geraldine Laybourne [Oxygen]

networks would just keep growing. Finally the arguments in favor of pay-television were beginning to prove themselves.

The spectrum of programming certainly had changed and expanded during satellite cable's second decade. The several dozen satellite basic and pay-cable services were diverse in terms of the range of audience constituencies addressed, the number of traditional television genres represented, and the number of new strategies being tried. Ironically, this broad range was the effect of increasing corporate consolidation (representing both horizontal and vertical integration) within the cable industry even more than it was an effect of entrepreneurial ingenuity. The more consolidated the industry became, the wealthier the successful entrenched entities became – and the more it was possible for existing networks (and their parent corporations) to risk new ventures. To gauge the extent of this consolidation, one only has to be aware that most satellite networks were owned, in whole or in part, by one or more cable MSOs (or other media conglomerates with MSOs among their holdings) as of 1992. Corporate giants like TCI not only were gobbling up smaller MSOs and independent local cable systems, they were also assuming significant ownership stakes in cable networks. Even Ted Turner's formidable programming empire was not immune. In 1986, following his awkward acquisition of MGM, Turner found himself financially overextended and at risk for losing control of CNN to a hostile takeover. It took intervention by John Malone – in the form of a debt restructuring that involved several cable MSOs – to save the company. Of course, the result was partial ownership of the Turner networks by the MSOs involved, not least of all TCI (Robichaux: 90–1).

Cable MSOs, TCI especially, were also able to prevent non-MSO owners of cable networks from gaining too much control of the industry. Robichaux relates the story of how, in 1984, ESPN tried to add a subscriber fee of 25 cents (to be increased to 30 cents after one year). The sports network, recently acquired by ABC, was well aware of the role it played in selling cable service. But Malone came up with a devious counteroffer. With backing from major sports advertiser Anheuser-Busch, Malone announced a competing network, Sports Time. ESPN immediately agreed to a compensation agreement more favorable to TCI. That same year, MTV tried a similar maneuver, and Malone fired back in similar fashion – this time in the form of backing for a music network proposed by Ted Turner (87–8). Indeed, TCI and other MSOs were able to negotiate volume discounts with nearly all cable networks. If it was not always necessary to threaten such cutthroat measures as just described, the mere possibility of those measures did much to maintain their competitive advantage. Virtually none of the savings were passed on to cable consumers, though, leaving them with subscription rates that seemed to be increasing exponentially.

The 1992 Cable Act

This apparent abuse of power presented a dilemma for federal policymakers, especially those who opposed a laissez-faire approach to the media and telecommunications industries. So, after years of dealing with the results of the highly deregulatory 1984 Cable Act (and the deregulatory policy decisions that had preceded it), Congress decided it was time to reinstate some curbs on the cable industry's expansion. This time they were trying to balance the interests of the NCTA with complaints from such disparate parties as the National Association of Broadcasters (NAB), the FCC, various consumer and citizen organizations, and rural constituencies (C-band dish owners and others lacking access to cable service). As discussed in earlier chapters, broadcasters had long been unhappy that, as they felt, cable operators were taking their signals without payment and then making a profit from those signals (broadcast signals still were, at this stage, the main selling feature of cable service). Although the 1976 Copyright Act had addressed these concerns to some extent, a need for more compensation was felt as cable's fortunes continued to accrue. And certain other telecommunications industries, realizing that cable would be a long-term competitor, sought provisions to guard their own interests. Notable among these industries were the various new telephone companies that had resulted from the federal government's 1984 break-up of the AT&T monopoly. The new DTH satellite industry, growing ever more confident of its market position, also sought measures to level the multichannel television playing field.

The new Cable Act (formally named the Cable Consumer Protection and Competition Act) was passed in fall of 1992 and reflected similar shifts in political sentiment to those that put Democrat Bill Clinton in the White House very close to the same time. In fact, the Act was spearheaded in large part by Senator Albert Gore (D–Tennessee), by then Clinton's running mate. Most in the cable industry held Gore in outright contempt, some even referring to him as "Darth Vader." The other major lawmaker behind the new Act was Senator Daniel Inouye (D–Hawaii), known for his efforts to keep big business in check. In the months prior to the presidential election, Gore divided his time between the campaign trail and shepherding the bill through to its passage – a process that included voting to override a veto by President George Bush. It was an issue Gore felt very strongly about, seeing US consumers as the victims of a greedy and out-of-control cable industry that had been allowed free rein by government policymakers. He was quoted as saying at the time: "Big cable companies, with President Bush's permission, have raised rates faster than inflation, shut out competition, and answered to no one but themselves since the cable industry was deregulated [in 1984]" (Camire).

Provisions of the 1992 Act were true to its title. One of its major sections sought to help consumers by reinstating different forms of rate regulation (to be determined by service tier). Other provisions sought to equalize competition between cable and its old and new rivals – broadcast television and direct broadcast satellite (the large C-band dishes in use at the time as well as the smaller Ku-band dishes that were just starting to be marketed). Probably the most memorable and precedent-setting provisions of the Act were the ones known as "must-carry" and "retransmission consent." These relate to the fact that certain broadcast stations are more popular than others; while some (e.g., major-market network affiliate stations) draw enough interest that cable systems would be willing to pay to carry them, others (e.g., independent stations with small programming budgets) would prefer to require systems to carry them as a way of broadening their audience. The Act thus required that cable systems offer local stations a choice: mandatory carriage, an option desirable for less popular stations, *or* compensation (financial or otherwise) for the use of their signals, the likely choice for more popular stations (in other words, those stations that cable systems needed in order to draw subscribers).

There were several other provisions to the 1992 Act that are worth noting. For instance, the cable industry felt the impact of a provision limiting the total number of subscribers to be reached by a single MSO as well as one limiting cross-ownership between cable systems and other satellite-based technologies within a given market. Other provisions addressed aspects of the local franchising process. And there were several provisions aimed at protecting consumers from unreasonable billing practices, invasion of privacy in the record-keeping and billing processes, and quality of signal. Some of these provisions would endure, others would prove too difficult to enforce, and others simply became anachronistic as new technologies reshaped the multichannel television landscape.

The State of Cable Technology

The trouble with implementing any cable policy, of course, is the rapidly changing nature of the industry and the technology it relies upon. Politics aside, one reason why the 1992 Cable Act followed so closely on the heels of the 1984 Act is that no one could have predicted with any degree of accuracy how quickly and to what extent the very nature of cable service would change in less than a decade. The role of technological change would be even more apparent in the much more comprehensive 1996 Communications Act – as will be discussed in the next chapter. Meanwhile, the conceptual groundwork for critical new technologies for the late 1990s and early 2000s was already being established in the 1980s and early 1990s.

The much-heralded first experiment with interactive cable, Warner-Amex's QUBE system, had been discontinued (though not without some revealing discoveries about consumers' expectations for cable service, including a desire for greater interactivity in home shopping), but the notion of interactivity had hardly disappeared from the cable industry's radar. With no concept of the Internet yet in the public mind, the cable industry continued to promote itself as poised and ready to fulfill science fiction-like visions of consumer-oriented inter-active audiovisual services. Already, home shopping services were extending the rudimentary interactivity of infomercials in order to fill whole programming days. Pay-per-view (PPV) channels were also on the rise, offering sports and concerts in addition to movies. In both cases, the small degree of interactivity achieved by a combination of on-screen menus and telephone-based ordering systems hinted at what was to come in the next decade, when computer-equipped set-top cable boxes would be introduced on a large scale.

Another innovation that helped cable networks distinguish them-selves from broadcast television during the 1980s involved the increased use of the *sideband* portion of a network's signal to enhance and update that network's content. The process of modulating a signal on the electromagnetic spectrum allows for some lower-grade signal to be available either above or below the main signal. While not techni-cally suitable for transmitting sound or moving images, the sideband can be used for transmitting electronically generated characters. Thus networks were able to introduce continuous updates of breaking news, weather, sports scores, stock prices, and so on – particularly beneficial to such early cable networks as CNN, the Weather Channel, ESPN, and CNBC that were competing with broadcast network and local station news.

The period from 1980 to 1992 also saw some of the earliest changes in cable system architecture (i.e., the wiring configuration that carries signals from the headend to subscribers' homes). Local systems, many of whose franchises were expiring toward the late 1980s or early 1990s, were promising extensive system rebuilds – often involving the replacement of coaxial cable with fiber-optic cable. This of course ushered in other new system components, particularly those related to digital transmission. And with the costs of this new technology came even greater efforts to centralize operations into large metropolitan "hubs" (or regional headends) that would use "spokes" of trunk line to connect to existing local cable systems, as mentioned above.

A cable system's technological infrastructure represents a great deal of equipment and construction expense – sometimes extending into the billions of dollars. So it would not make sense to replace an older plant entirely during a rebuild. Instead, new components are integrated with older ones. For example, many systems have initially installed fiber-optic cable only for the trunk lines, with fiber-optic feeders and taps to come later. Similarly, set-top boxes seldom are replaced all at once when newer models become available. Instead they are introduced with new subscriptions, changed at subscribers' request, or brought in as replacements for older boxes as they wear out.

With all the old and new equipment that had begun to be interconnected during the 1980s, and with the many new entities developing cable-related technologies (not to mention technologies such as computers that previously had barely been considered cable-related), it became very obvious that more attention needed to be paid to ensuring the interoperability of the various components. The NCTA, of course, played a major role in standardizing cable technology, as did various cable trade organizations (particularly the Society for Cable Television Engineers, or SCTE) and a new meta-industry research and development firm called CableLabs that would help operators and equipment manufacturers develop and implement common standards.

In 1984, electronics engineer and cable operator Richard S. Leghorn perceived that the cable industry's technological development had been left almost entirely to the proprietary interests of individual component manufacturers. That industry was moving ever farther from a future in which individual components made by different companies would be compatible with one another. Leghorn therefore proposed to the NCTA the development of a research and development entity that would employ "systems engineering" or "managed system development" techniques to develop industry-wide technical standards. Some four years later, in 1988, this entity, called Cable Television Laboratories, Inc. or simply CableLabs, was founded and eventually located in Colorado, near the growing Denver-area hub of cable companies. CableLabs maintains an important role as more and more forms of electronic technology become integrated with the traditional functions of cable television.

The Synergy Wave Begins: Mergers and Acquisitions

The late 1990s marked the start of an era in which the cable *business* – along with the technology it relied upon – would grow less and less distinguishable from related media. Nowhere was this shift more apparent than in the 1989 merger of Time, Inc. and Warner Communications – both already among the largest cable programmers and MSOs – to form Time Warner. The new media conglomerate would not only become a behemoth in its own right, initiating further mergers and acquisitions as well as launching new media businesses; it would set in place a corporate environment in which the only means of competing would be to merge with other large companies, acquire smaller companies, and develop new ventures (especially ones involving cutting-edge technologies). Of course, this was a time when new media technologies and accompanying business ventures were being introduced to consumers nearly every day. Within a decade, cable as a standalone medium would seem relatively antiquated and those in the cable industry would be racing to find new and related sources of revenue in order to stay competitive (for more on the Time-Warner merger, see Box 1.1).

Other mergers had already occurred during this decade that, while not yet involving cable and satellite television directly, would have powerful impacts on the overall shaping of television's multichannel era. One was the 1986 acquisition of the ABC broadcast network by Capital Cities Broadcasting, owner of multiple broadcast television stations. Not only did the newly merged company control what was still one of the three major television programming outlets in the United States, it also controlled all of the former "CapCities" stations plus ABC's prestigious "O & O" (owned and operated) stations in major urban markets including New York, Chicago, and Los Angeles. At the same time, Australian media mogul Rupert Murdoch was in the process of becoming a US citizen and acquiring the six-station Metromedia group that would form the basis for the new Fox network, US television's fourth broadcast television network. Also, in 1989, Japanese electronics corporation Sony acquired control of the Columbia Pictures and TriStar movie studios – recall that TriStar had been created largely as a means of channeling Hollywood movies to HBO.

The Rise of Multichannel Competition

Meanwhile the cable industry also was facing competition in the very areas of service that had come to define it. First, for a brief period around 1980, broadcast forms of pay-television made a comeback in large cities. Notable systems included WWHT, Wometco Home Theatre, in metropolitan New Jersey, and KBSC, National Subscription Television, in Los Angeles. There were a handful of others around the country. The broadcast pay systems enjoyed some brief success during the years when larger cities had not yet been wired for cable. But these one-channel systems, which needed only a broadcast frequency and transmission equipment to operate, were quickly supplanted by multichannel cable once it arrived.

Another form of television that relied on the airwaves for transmission was MMDS (multichannel multipoint distribution system), which used microwave relays to deliver up to ten satellite cable networks from a central transmitter within a limited area (microwave, being a line-of-sight form of transmission, cannot extend very far without an extensive series of retransmitting towers). The first company to offer MMDS service was Microband Corp., starting in 1982. MMDS continued in the United States, with limited success, from the mid-1980s through the late 1990s (and is still in use in other countries). In its early years, subscription and equipment costs for MMDS made it a reasonable alternative to cable under some circumstances, but the more channels that became available through cable and other technologies, the less desirable MMDS service proved to be for consumers.

More competition for the cable industry came from SMATV (satellite master antenna television), which uses a headend-type rooftop receiving dish to deliver cable networks' signals to hotel guests or residents of large apartment buildings. SMATV companies have thrived in urban areas, where they can serve multiple customers without needing to secure permission to use public rights of way (e.g., stringing wires on utility poles, running wires across streets). Naturally the cable companies serving areas where AMATV systems exist have been unhappy about the competing technology, primarily because it has made them somewhat less successful in convincing landlords to provide access to buildings where tenants wish to subscribe. SMATV companies typically split revenues with landlords, and thus have had little difficulty gaining access. Eventually cable companies began to work out similar agreements – or even entered the SMATV business themselves. The cable companies won their battle with SMATV in large part because of the extreme pressure they were able to put on the programming networks; it would take more SMATV systems than even existed to generate the subscriber fee revenues earned by bulk contracts with the large and powerful cable MSOs.

The most serious competition for the cable industry of the 1980s came from the home satellite (DTH or direct-to-home) industry – something of a regulatory "no man's land" in multichannel television. During the late 1970s the cost of the C-band satellite receiving dishes used at cable headends fell to around $2,000–$3,000, low enough to be purchased directly by consumers. This was a benefit to isolated rural residents, a great many of whom had been deprived of both broadcast and cable television service. Most of these households had the yard space needed for the 10–12-foot C-band dishes. For a few years in the early 1980s, the one-time cost of the dish and related amplifying equipment were all that was needed to receive cable networks' signals – in fact, an array of networks wider than any cable operator was able to provide at the time. With costs spread over a few years, this proved a bargain compared to cable service. Before long, consumers who *did* have cable service available where they lived began using the DTH technology in order to avoid cable fees.

A few "do-it-yourself" books were published, representing a small but vocal minority among television viewers. These books, which represented a short-lived trend, encouraged consumers to bypass the supposedly restrictive world of cable service altogether. They tended to use utopian language similar to that of the Blue Sky documents. Mark Long, for example, wrote: "Perhaps you feel that receiving satellite TV is like booking a seat on the space shuttle – too daring and complicated an adventure for you at this moment in time. Not true. There may be a few more knobs and buttons, but it's not really that much different from your regular TV" (9–10).

It did not take the cable industry long to become aware of – and very concerned about – the lost revenues (as well as potential revenues from rural areas that operators hoped to serve in the future). The home dishes drew the attention of cable programmers in much the same way. They were missing out on subscription payments for premium channels and per-subscriber fees for basic networks (at this early stage, measuring lost advertising revenues was not as much of a concern, though it would be in the near future). Both groups, largely through the NCTA and other trade organizations, began to pressure policy-makers for relief. They also began to pursue technological solutions to the problem: most prominently, signal scrambling. With scrambling, DTH users would have to pay for the services they wanted through an intermediary business similar to a cable operator.

The 1984 Cable Act helped this along. It gave cable programmers a lot of incentive to begin scrambling their signals. The Act formally permitted dish owners to receive signals – but only provided those signals were not scrambled and that a marketing plan was not in place whereby subscribers would have had to pay for cable-like programming packages. For a brief time, this created a more level playing field – allowing those without access to cable to receive cable-type programming packages, but generally discouraging those in cable service areas from putting up the clumsy and expensive dishes.

In 1985, HBO became the first cable network actually to scramble its signals on a regular basis (with others soon to follow). At that point, even though backyard dish ownership had grown to two million, sales suddenly plummeted. Irate dish owners lashed out when they learned they would have to pay for cable-type subscriptions to both premium and bundled basic cable networks. In one notorious and extreme case, John R. MacDougall, a dish salesman moonlighting as an engineer at a satellite uplink facility, briefly replaced HBO's signal with the following teletext message:

GOOD EVENING HBO FROM CAPTAIN MIDNIGHT
$12.95/MONTH?
NO WAY!
(SHOWTIME/MOVIE CHANNEL BEWARE)

MacDougall's eventual guilty plea in this misdeed brought him a fine of $1,000. The cable industry learned a more important lesson: that it was subject to competition brought on by the very same technologies it relied on to increase its own business. This would only become more apparent in the later 1990s, once the much smaller Ku-band DTH dishes reached the market.

Meanwhile, one of cable's future competitors, the Internet, was in its embryonic stages – nowhere near ready for use by the general public, but slowly coalescing into the power communication tool it would be in the future. The Internet's origins lie in a network of computers known as ARPANet (Advanced Research Projects Agency Network). ARPANet was first developed by the US Department of Defense in the 1960s (the height of the Cold War) as a decentralized system of communication to be used in the event of one or more nuclear strikes. With no such attacks occurring, however, the network was eventually adopted and adapted for use by large universities and other research centers (including the Department of Defense itself). ARPANet was sluggish at this stage, and its use limited to large mainframe computers with similar information transfer protocols (i.e., information-sharing procedures). It was not until the late 1970s that even the most rudimentary forms of the TCP/IP protocol – a computer code that would enable everyday consumers to access the network – were developed. But by the early 1990s, TCP/IP protocols had been refined, allowing users to access the Internet via many different platforms (i.e., types of computers and operating systems). At this stage, Internet users needed to be familiar with the UNIX computer language in order to use the Internet effectively, but by the middle of the decade, user-friendly browser programs such as Mosaic, and later Netscape, would become available. From that point, it would be only a matter of months before personal, institutional, and commercial websites began to replace the experimental sites that had been developed by computing professionals to test the new technology.

It should not come as a surprise that cable operators and related telecommunications businesses were quick to perceive the potential of the Internet once it was embraced by the general population. It also should not be surprising, either, that media producers and distributors became very interested in the array of digitized information showcased on the Internet. By the early 1990s, some were already beginning to experiment with digital information and entertainment products themselves – as will be discussed more in Chapter 7. As of 1992, however, cable was still dominant among the various multichannel television providers.

The International Scene

American-style cable was also increasing its hegemony around the world. The 1980s were a decade during which most traditional public service broadcasters throughout the world began to feel strong pressure from both national and transnational private program services – both terrestrial and satellite-based – that echoed the business strategies of US cable networks. Most of the public service broadcasting organizations had been founded during the 1950s or 1960s in order to advance social goals such as decolonization, economic improvement, and public education. Their programming sometimes was considered low-budget and rather boring, if not outright didactic or propagandistic. But as state-funded monopolies, they saw little reason to change. By the mid-1980s, however, things were different. As national governments made ever greater efforts to expand the quality and scope of their broadcasting services, operating costs skyrocketed, and the lure of advertising dollars to supplement public funds became irresistible. Some countries, including Canada and Germany, had been balancing public and private funding since the beginning; this practice only spread – particularly into developing countries – during the 1980s.

As another example, India's socialistic national broadcaster, Doordarshan (DD), had been in the business of creating a national identity and common set of economic development goals among disparate social classes as well as urban and rural residents. But with efforts to expand the television infrastructure into rural areas and to serve more ethnic minority populations, operational costs went up dramatically – leading to a need for advertiser support. Amos Owen Thomas explains:

Despite programming inadequacies, DD penetration continued to grow, because between 1982 and 1992 the number of government television transmitters increased from 39 to 531. This emphasis on hardware was criticised for having led to dependence on advertising revenue and neglect of programming quality. Over the same period the number of television sets also increased significantly from 2.1 million to an estimated 34.9 million sets, making for a maximum potential Indian audience of potentially over 200 million individuals. (100–1)

By the 1990s, once India had made a definitive shift from a socialistic to a capitalistic economic system, residents gained access to private satellite broadcasters, including StarTV (transnational) and ZeeTV (national). DD itself responded by creating subservices DD1 through 6, which targeted both regionally and by programming category (Thomas: 102).

In the mid-1980s, the Mexican government nationalized the domestic satellite industry. Mexico's powerful private broadcaster Televisa countered by establishing the international satellite system PanAmSat across the border in the United States, intending to capture viewers both in border states and in Mexico itself. Although Televisa had entered the US broadcasting market more than a decade earlier, this was even more of a definitive move toward regional dominance of the Spanish-language television market. By the early 1990s, Televisa would hold stakes in television broadcasters throughout Latin America.

Cable- and satellite-delivered multichannel television got off to a slower start in Western Europe. This was not a problem of government restrictions, as it was in other parts of the world. Rather it was due to the presence of a different sort of preexisting industry structure. In most Western European nations only one television channel was available, the publicly funded public service channel. So as of the 1980s, commercially funded *broadcast* alternatives (including digital broadcasts) were only starting to enter the picture. With these "free" channels offering a previously unknown degree of televisual choice, most Europeans saw little need to pay for even more options – at least not at this stage. Furthermore, it was more difficult to find the critical mass of subscribers needed to sustain pay services in Europe because the potential audience was so fragmented in terms of language and cultural traditions. The pay channels would arrive in greater numbers in the 1990s, though, helped by the growth of the

European Union, the accompanying proliferation of English as an official *lingua franca*, and the rise of capitalism in Eastern Europe after the fall of the iron curtain.

It is also worth pointing out that a rather different use of cable technology had begun in some nations by the 1980s: the wiring of urban neighborhoods or apartment blocks for the purpose of show-ing pirated movies and other material (including pornography). In the Soviet Union and other Soviet bloc nations, for example, this offered an alternative to the state-controlled broadcast television channels. In developing nations such as India, it was a low-cost alter-native to other forms of entertainment. India's *cablewallahs* repre-sented a sort of "gray market" – low-budget operations wherein a small entrepreneur such as a homemaker or student would set up a slapdash office-headend operation in order to provide programming to a limited area. Thomas explains:

> The average *cablewallah* operated from a small room in an apartment block on the roof of which there were two or three satellite dish antennae. Along one or more walls would be a bank of VCRs which taped satellite broadcasts for later transmission or played movie videotapes. Along with a divider, receiver, frequency modulator and booster, the total investment was about Rs 50,000 (US $1,700). Each cable operation covered a radius of half to one kilometer and every neighborhood had about four operators, each differentiating itself by service and programme offerings. (119)

This was not "cable television" as people in North America had come to know it – though the *cablewallah* operators, in their entrepre-neurial spirit, do seem reminiscent of the CATV operators in the United States of the 1950s. And like the CATV operations in the United States, *cablewallah* operations surely helped to plant expecta-tions for the Western-style commercial television that would arrive via satellite in most places during the 1990s.

By this point it seemed apparent that multichannel television was desired (even where not yet feasible) in nations throughout the world. However it was the concept – rather than a prescribed technological configuration – that was important. North Americans easily assume that "cable television" historically has been synonymous with multichannel television, and that other technologies such as DBS are simply late-arriving competitors. This chapter has provided some contrasting examples to this notion. The next chapter will provide more. It will also look briefly at the current media scenario, in which multinational capital increasingly challenges regional and national identities in defining televisual information and entertainment.

NOTE

1   At this early stage, there were some non-satellite-based regional services, some broadcast services relayed by cable systems, and some satellite services that shared a common transponder.

FURTHER READING

Auletta, Ken. *Media Man: Ted Turner's Improbable Empire*. New York: Norton, 2004.

Denisoff, R. Serge. *Inside MTV*. New Brunswick, NJ: Transaction Press, 1988.

Robichaux, Mark. *Cable Cowboy: John Malone and the Rise of the Modern Cable Business*. Hoboken, NJ: John Wiley, 2002.

Thomas, Amos Owen. *Imagi-Nations and Borderless Television: Media, Culture and Politics Across Asia*. Thousand Oaks, CA: Sage, 2005.

Chapter 7

Multichannel Television's Mature Years: 1993–Present

By the early 1990s, as cable television approached its fiftieth anniversary in North America, the industry must have seemed barely recognizable to those who had been involved since the beginning. The medium that for so long had stymied government policy-makers, to the point of virtual inaction, was now under their constant – if not entirely effective – scrutiny. And if only a handful of rural residents had been familiar with community antenna television in the 1950s, modern cable and its futuristic promise were on everyone's mind by the 1990s. Where once cable's greatest programming innovators had been those who figured out how to import popular stations from distant markets or air rudimentary local news on vacant channels, cable now boasted hundreds of its own programming networks – most with original programs for at least some portion of their schedules.

From this point, the major innovations in cable and direct broadcast satellite would have somewhat less to do with programming than with delivery technology and related services – even though programming innovation did continue apace. The 1990s and early 2000s have witnessed the rise of new pay-per-view systems, digital service tiers, and a host of other innovations to existing multichannel television service. This period has also seen the rise of the Internet – and its groundbreaking data transfer protocols – from an obscure mode of communication among university and government researchers to a popular, multifaceted, and heavily commercialized communication medium. The cable industry would play a key role here as well, offering consumers high-speed access to the increasingly data-heavy (and therefore slow transferring) array of Internet content.

Government policy-makers, clearly reeling from the speed at which media industries and technologies were changing, rushed to pass a major body of legislation meant to address a media scenario never imagined by the writers of the 1934 Communications Act (still the most recent comprehensive media-related legislation to be passed). Unlike the medium-specific Cable Acts of 1984 and 1992, the 1996 Communications Act was intended to address a rapidly *converging* array of media – not to mention the many corporations vying for control of those media.

And while this was taking place in the United States, other nations from all corners of the world, along with an increasing number of transnational media entities, were joining this information "revolution." Nations such as those in Western Europe and the British Commonwealth, with longstanding public service broadcasting operations, had clearly moved into the multichannel television era by this stage, with the older public channels now accompanied by (and competing with) a range of national or international private channels available via traditional terrestrial broadcast, cable, or direct broadcast satellite. Residents of other nations, particularly those in rural or impoverished parts of the developing world, were watching television for the first time ever, since satellites allowed television signals to reach areas not served by broadcast stations or cable lines. New technologies also brought the mixed blessing of "transborder" television, whereby programming produced by larger or wealthier nations easily reaches viewers elsewhere, introducing concerns about cultural imperialism on an entirely new scale. Where previously program importation had been under at least the nominal scrutiny of government authorities, satellites have made these information flows much harder to control. In fact, major changes such as the fall of the iron curtain and the rise of China as a major world economic power have been at least partially attributable to shifts in available media. Clearly global television since the early 1990s has been both cause and effect of major political, cultural, and economic changes.

The Industry Structure

Throughout the 1990s, it continued to be "business as usual" for larger cable MSOs such as Time Warner, TCI, and Comcast to buy out smaller MSOs and a number of the remaining independent systems. Some of the more prominent acquisitions include Time Warner's purchase of NewChannels in 1995, US West's acquisition of Continental Cable (forming the subsidiary MediaOne, which was then spun off as a separate entity), and AT&T's acquisition of MediaOne along with most of TCI's systems in 1998 (and Comcast's subsequent acquisition of many of those systems in 2001 through its merger with AT&T Broadband). The hub and spoke architecture of most new cable systems also led to regional consolidation of operations. For example, a small city such as Racine, Wisconsin (population 92,000) – which as recently as the late 1990s was owned by TCI and had its own local headend – is now being served out of Time Warner's Milwaukee hub, some 40 miles away. This was the result of an agreement in late 1998 between Time Warner (with around 12.6 million subscribers at the time) and TCI (with around 15.75 million) to form metropolitan hubs: greater Milwaukee for Time Warner and greater Chicago for TCI. Other metropolitan areas affected by this wave of system "swaps" include Portland, Oregon; Portland, Maine; and St. Louis, Missouri (Dempsey). Thus in 1997, for the first time in history, the number of local cable franchises in the nation actually dropped – a pattern that would continue (Compaine and Gomery: 251).

After 2000, following the breakup of TCI, Comcast would grow into the new giant MSO, boasting 23.3 million subscribers as of 2006, compared to second-place Time Warner's 14.5 million in the same year. Third place Cox Communications had around 6.3 million, but number ten, CableOne, only had around 700,000. Service Electric, the company John Walson claimed to have started in 1948, today is ranked at number fifteen, with approximately 300,000 – most within 100 miles of Mahanoy City (Hearn and Farrell, NCTA Cable Developments). Smaller MSOs still exist, as do several small, independent systems, but they tend to serve rural areas and communities in economic decline – places that would only be desirable service areas for larger MSOs if they could easily be integrated into existing urban hubs.

Another significant trend among cable MSOs that gathered steam in the 1990s has involved efforts to merge cable businesses – operators as well as programmers – with other media and telecommunications businesses. This synergy trend really began with the 1989 merger that formed the colossal Time Warner conglomerate [see Box 1.1].

During the 1990s, Time Warner not only acquired a number of local cable systems around the country, through both acquisitions and swaps, it also increased its programming holdings as well. Most noteworthy, the company merged with Turner Broadcasting in 1996, adding several of cable's most popular networks to its already synergistic holdings. As Ken Auletta explains, "If Time Warner owned TNT, the Cartoon Network, or CNN, it could generate advertising revenues from the programs as well as subscriber income from monthly cable users, and gain leverage over other cable programmers since Time Warner could threaten to favor its own cable networks with lower and more desirable cable channel slots" (68).

If the $183-billion merger that had formed Time Warner in 1989 had seemed colossal at the time, the merged company's acquisition by the online service provider America Online (AOL), finalized in early 2001, made the conglomerate even larger and more synergistic still. Not only did this merger represent extensive cross-promotional opportunities (e.g., websites to promote movies and television programs), it was also a move toward a future scenario in which more and more audiovisual entertainment would be delivered to consumers via the Internet. In fact, Time Warner has been among those most actively pursuing online program delivery strategies and modes of funding in the twenty-first century.[1]

The 1990s and early 2000s saw other, similar corporate mergers and acquisitions that would also go far toward shaping the multichannel television environment of the twenty-first century:

- 1993: Walt Disney Co. acquisition of Capital Cities/ABC
- 1997: Microsoft acquisition of 11.5 percent of Comcast Corp.
- 1997: News Corp. acquisition of International Family Entertainment (owner of the Family Channel and television producer and syndicator MTM Entertainment); Family Channel renamed Fox Family
- 1999: CBS and Viacom merger
- 2001: Disney acquisition of Fox Family, subsequently renamed ABC Family

Smaller, though nonetheless significant mergers and acquisitions during this period included several purchases of broadcast radio and television station chains, purchases of cable MSOs by other MSOs as well as by telephone companies outside their own service areas (newly allowed by the 1996 Telecommunications Act), and purchases of Internet companies by both other Internet companies and more established media conglomerates.[2]

Aftermath of the 1992 Cable Act

Some of the mergers and acquisitions taking place during the mid-1990s were fairly direct (if somewhat unintended) results of the 1992 Cable Act. One of the two most significant outcomes of the Act involved the must-carry and retransmission consent provisions. The FCC, urged on by a powerful NAB, had continued to express concern about the effect of the new cable television industry (still facing very little multichannel competition at this stage) on the established broadcast television industry. So the must-carry provisions of the Act (recalling FCC regulations of the mid-1960s) were aimed at forcing cable systems to carry local broadcast signals ("local" being defined as a television station's service area). The major beneficiaries of this provision were small network affiliate stations facing potential competition from larger affiliates within the region (for example, a Harrisburg or Altoona, Pennsylvania station competing with Philadelphia or Pittsburgh stations) and small, independent UHF stations (often home shopping or religiously affiliated stations) for whom cable carriage represented a substantial increase in coverage area.

However, not all broadcast stations needed the must-carry provisions to preserve their fortunes in the multichannel era. Network affiliates in larger urban markets have remained major providers of Americans' television programming, particularly during the lucrative prime time hours. Thus they benefited more (some would argue, extravagantly) from the Act's retransmission consent provisions, which stations were permitted to choose instead of must-carry. Retransmission consent allows a station to request compensation, financial or otherwise, for the use of its signal by a local cable system. And since network affiliate stations have represented the lifeblood of cable television, refusing to provide this compensation could have put a cable system out of business. Not all broadcast stations actually sought financial compensation for the use of their signals, however. Instead, quite a few chose compensation in the form of carriage of a start-up cable network in which they held financial interest. For this reason, a number of niche networks were launched successfully in the mid-1990s that might not have succeeded otherwise.

For example, the Scripps Company, owner of newspapers as well as a chain of television stations, was able to leverage the retransmission consent provisions into carriage for its new cable network Home and Garden Television (HGTV). And the Providence Journal Company (also a combined newspaper and television station owner) launched Television Food Network (TVFN, now called Food Network) in 1993. The Providence Journal Company was subsequently acquired by the Belo Corporation (also a newspaper and television station owner). Scripps purchased Food Network from Belo in 1997. Scripps has since parlayed its success with these "lifestyle" networks into a cable programming empire that also includes the DIY and Fine Living networks. The retransmission consent provision of the 1992 Act also helped boost carriage of cable networks owned (or partly owned) by broadcast *networks*, including Capital Cities/ABC's spin-off, ESPN2.

The other major outcome of the 1992 Act, rate regulation, epitomizes the concept of "best-laid plans gone awry." Politicians and policy-makers, though driven by the most consumer-minded of intentions, proved extremely uninformed about how the cable business actually worked. In fact, cable's established ways of doing business in the satellite era severely undermined the Act's benefits to consumers. As Mark Robichaux argues:

> [S]ubscribers didn't get much of a break. Many systems created new packaging schemes to avoid regulation. Most had been charging higher rates than the FCC benchmarks would allow for the most popular, expanded tiers of service. But their prices were far below the FCC limits for basic cable packages. So the cable operators reduced charges for the costlier packages and recouped the money by raising basic cable rates to the cap set by law. That shift defied a major goal of the new law: to safeguard the affordability of basic TV service for low-income people. (128)

Relations between the cable industry and US government policy-makers only worsened from this point, especially as cable industry rival Al Gore assumed his vice-presidential role and the overall US political climate shifted to a somewhat more regulatory one than had predominated during the Reagan–Bush years. Moreover, it quickly became clear that the 1992 Act (or any other legislation) had done virtually nothing to address shifts to the US media scenario writ large – including cable along with many other technologies and industries that were growing increasingly intertwined in their operations.

The 1996 Telecommunications Act

In other words, the almost science fiction-like convergence of media technologies that clearly was in process by the early 1990s brought with it a widespread sense that various industry-specific regulations and legislation enacted since passage of the 1934 Communications Act were no longer sufficient in and of themselves to keep that encompassing Act relevant. Another sweeping piece of legislation would be necessary to address the economic and technological ramifications of media technologies – including computers, fiber-optic cable, and satellites – that had been unknown to policy-makers of the early 1930s. So, in accord with the overall tenor of the Clinton era, the 1996 Telecommunications Act was drafted with the goal of maintaining regulated free-market competition in the converging media industries.

The Act embodied divisive issues, though not always cast along traditional party lines. Supporters included not only old-style conservatives and proponents of laissez-faire economics, but also more moderate liberals (as represented by the Clinton–Gore administration) who believed that traditional regulatory frameworks were no longer applicable in an era of synergistic media conglomerates, digitized information, and converging media technologies. Patricia Aufderheide explains that the blurring of the traditional roles of media companies "was now seen as in the public interest, meaning promotion of innovation and a marketplace that could foster a greater range of products. That could even mean giving incumbents an advantage" (30). By "incumbents," Aufderheide of course is referring to gigantic, entrenched, and often synergistic media or telecommunications corporations ranging from TCI to Time Warner to AT&T.

Opponents to the legislation included, not surprisingly, consumer and public interest advocates with legitimate concerns about monopolistic or oligopolistic norms such as price-fixing, failure to subsidize service to low-income and rural consumers, limitations to First

Amendment rights, and diminished quality of service overall. While on the one hand deregulation seemed to be increasing the diversity of telecommunications services available to the public and lowering prices, on the other hand these could be seen as short-term results. In a deregulated media scenario, smaller companies tend to be bought out by larger companies at the same time as large companies are merging with other large companies.

More than a decade after passage of the Act, the media consolidation issue is still hotly debated. The key players have shifted due to mergers and acquisitions, but typically this is done to trade out older businesses for newer, more lucrative ones. It is difficult for policy-makers to keep this under control. For example, the availability of streamed versions of TV shows and movies over the Internet surely has had a mitigating influence on their concerns about a monopolistic cable industry. In fact, the Internet may well have replaced DTH satellite service as the most prominent aspirant to cable's fortunes. But cable companies have been no less involved in delivering content through these new technologies than they were with the earlier technologies.

The Act made some gestures toward consumer protection, with nods toward such vaguely articulated concerns as "universal service" and "fair pricing." It also provided for public services, including Internet access for K-12 schools and public libraries (though not to other public entities such as colleges, community centers, or cultural organizations). And it tried (though with very limited success) to address child protection concerns voiced by citizen-consumers. It mandated that the V-chip, a device to allow the blocking of un-wanted programs, be installed in new television sets. It also mandated a program-rating system for broadcast and cable television, to be used by parents with or without the accompaniment of the V-chip. And the "communications decency" provisions made it a criminal offense to knowingly distribute "indecent" material to minors via the Internet. This last is extremely difficult to define, much less enforce – and was eventually ruled unconstitutional. The traditional "community stand-ards" gauge of decency hardly applies to a medium that reaches every part of the world, and US authorities have no jurisdiction over mater-ial originating in other countries. It also has been widely challenged on First Amendment grounds.

But the real and lasting impact of the 1996 Telecommunica-tions Act is in the foundation it laid for the media industry structure heading into the twenty-first century. Whether intentionally or not, the 1996 Communications Act effectively set the stage for a contest over "ownership" of the digital information concept – a contest involving such players as cable operators, telephone companies; Internet service providers (ISPs), and a host of other digital content owners, producers, or distributors. The Act opened up areas where competition previously had been prohibited – for example, permitting (with limited restrictions) the "baby Bell" phone companies to offer long distance service and long distance carriers such as AT&T to offer local service. The Act also further loosened guidelines for areas in which competition already existed. It increased the percentage of the total population that may be served by a single broadcast television station owner (to 35 percent) and allowed virtually unlimited ownership of broadcast radio stations by a single entity. It also increased the license term for both radio and television stations from five to eight years, and indicated that existing owners would be strongly favored over any competitors at license renewal time. The Act offered no guidelines to the FCC for determining the meaning of "public interest" in these situations, however – an area of concern for media reformers dating back to the 1920s and the first efforts at media legislation.

This loosening of the reins was ostensibly intended to encourage the adoption of new technologies and business plans that would benefit consumers. For similar reasons, the Act provided for the giving away of valuable spectrum assignments to established broadcast interests in order to encourage the development of high-definition television (HDTV) channels. The implementation of this provision has been fraught with controversy and complications. Finally, the Act allowed common carriers, most specifically telephone companies, to begin offering video services – as will be discussed below.

Marketing the Cable "Brand"

As media technologies converged, helped along by the 1996 Act, media content followed. The technological and business convergences addressed by the legislation have been apparent in the operations of cable *programmers* as well as operators during the late twentieth and early twenty-first centuries. In the early 1990s, the trend of "name-brand" or "signature" cable programming initiated by early networks such as HBO, CNN, and MTV was only heightened by the advent of specialty channels such as Food Network, HGTV, the Discovery networks, and others. In fact, the medium of cable was coming into its own around the same time as the rise of segmented, lifestyle-oriented and "experiential" marketing. Although there are many different orientations here – most accompanied by popular business strategy manuals – the overall emphasis has been to target consumers according to lifestyle categories, as predicted by empirical measures such as the Values and Lifestyles (VALS) categories or the Claritas PRIZM clusters. As B. Joseph Pine II and James H. Gilmore explain in their popular marketing strategy book, *The Experience Economy*, "Only when companies constitute [information] in the form of information *services* – or informational *goods* and informing *experiences* – do they create economic value" (ix–x). In other words, engaging a potential consumer in an interactive, product-linked information environment is now believed to be an extremely effective way of selling products and services – especially those involving discretionary income.

Figure 7.1 USA's successful detective series *Monk* (2002–), featuring an "obsessive-compulsive detective" (Tony Shalhoub), is an example of the basic cable network's "Characters Welcome" branding strategy [The Photo Desk]

Programmers and advertisers quickly became aware that lifestyle-oriented cable networks suited this marketing function ideally. Just consider how much greater an impact commercials for building outlets such as Home Depot and Loew's would have when placed in a remodeling program on HGTV than they would on a widely targeted broadcast primetime program. And so successful has Food Network's star chef Emeril Lagassé been at cultivating gourmandism that he has been able to spin off his own line of products for aspiring home chefs. As these and other networks have developed coordinating interactive websites and program-linked product lines, their "experiential" pitches have grown all the more effective.

Lifestyle networks such as those just discussed also have expanded their brand-name niches outward generically. Food Network, for example, has moved into tangentially related program genres such as game shows (e.g., *Iron Chef America*) and travel shows (e.g., *$40 a Day*) – all with lucrative product and service tie-ins. At the same time, more established networks such as Viacom's MTV and Nickelodeon have grown much more aggressive in their experiential branding efforts. Nickelodeon has leveraged its child- and parent-friendly brand into an array of toys, clothing, theatrical movies, home entertainment products, and most recently theme-park attractions. It also has been one of the more pioneering cable networks in extending its televised programs into cyberspace. Episodes of its popular "tween"-targeted show, *Zoey 101*, can now be purchased and downloaded from Apple's iTunes store. And some newer television series are enhanced by backstory material available exclusively on the network's nick.com website.

The ESPN networks have taken an intriguing approach to branding their program "products" as well, by expanding the array of what constitutes a competitive (and television-friendly) sport. ESPN2 debuted in 1993 as a home for sports appealing to Generation X audiences and included such "extreme" sports as freestyle skiing, snowboarding, and skateboarding. The original ESPN network added competitive poker to its coverage several years ago (and spawned a range of poker-related programming on several other broadcast and cable networks) and now offers a number of tie-ins (including links to online poker) on its website. Still in search of other niches for drawing competitive sports enthusiasts, ESPN recently added competitive dominoes. As the *New York Times* reported:

The games almost always draw spectators, so perhaps it is no surprise that the ESPN sports network has declared dominoes the next big spectator sport and is promoting it as both a colorful cultural touchstone and a highly competitive game, complete with rankings, formal tournaments, celebrity events and sponsors.

Encouraged by the success of televised poker, the network has begun combing New York City for top players and colorful clubs for its coverage, and has been taping segments on formal tournaments and casual neighborhood games. (Kilgannon)

ESPN also has taken on Little League baseball and covers a portion of the annual Scripps National Spelling Bee.

Telephone Companies

Without a doubt, a trend toward interactivity and cross-platform marketing has been on the rise in cable programming since the mid-1990s. So it is not surprising that, starting in the 1990s, the US cable and satellite television industries faced new competition from the nation's telephone companies ("telcos"). The telcos, with their long history of providing two-way communications, saw themselves as natural players in the market for interactive televisual communications. But their enthusiasm for entering this arena also came as the result of a monumental shift in the regulatory climate governing them.

For just over a century, starting shortly after Alexander Graham Bell's invention and patenting of the telephone in 1876, the US telephone business had been controlled by a single entity, the American Telephone & Telecommunications Corporation (AT&T, casually referred to as "Ma Bell," after its founder). Anyone born before 1970 surely remembers the plain, large home phones that had to be rented from the AT&T monopoly. They surely also remember the high rates charged for long-distance service. All of this changed in 1984, at the height of the free market-oriented Reagan administration, when the US government finalized its breakup of the AT&T monopoly, dividing it into seven RBOCs (Regional Bell Operating Companies, also known as "baby Bells"). This move opened long-distance phone service to competition and allowed consumers to purchase and own their home telephones (bringing to market a plethora of designer models).

In the years to follow, not only did a number of long-distance carriers (including Sprint and MCI) begin competing head-to-head with AT&T; the RBOCs themselves entered a phase of merging with or acquiring one another and offering an array of new services.

The rise of new telephone technologies, most prominently mobile telephones, also played a significant role in the rapid changes to the telephone industry. Whereas in the early 1990s only a negligible percentage of the US population were mobile phone users, just one decade later there were clear indications that traditional landline telephone service was on the decline. In 2003, out of an average US household's annual telephony expenditures, $441 went to the local service provider (landline), $122 to the long-distance provider (landline), and $492 to a wireless provider. In 2005, there were 191.3 million mobile phone subscribers in the United States.[3] Traditional telephone companies' investment in mobile telephony thus has represented a tremendous revenue stream to these already wealthy corporations. Other telephony services such as caller ID, call forwarding, and voice mail have also helped the fortunes of the telcos since the late 1980s.

Throughout the 1990s and 2000s, the telephone companies have stood poised, ready, and eager to compete head-to-head with cable and DBS companies in delivering both video and high-speed Internet access to consumers. This (along with the added competition from DBS providers) created a high-pressure situation for the cable MSOs, who had grown accustomed to a virtual monopoly situation. As big as cable MSOs are, telcos are bigger still. For instance, for 2005, the newly merged AT&T (the name adopted by SBC Communications after acquiring the original AT&T) reported revenues of about $44 million, whereas cable giant Comcast reported only half that.

The final judgment in 1983 that broke up AT&T the following year had prohibited the RBOCs from offering what were vaguely termed "information services." The 1984 Cable Act then placed restrictions on telcos operating cable systems within their telephone service areas (though many launched or acquired systems elsewhere, particularly outside the United States). But prior to passage of the 1992 Cable Act, the FCC had already begun to reconsider the wisdom of this portion of the legislation – seeing telco entry into cable services as a means to promote healthy marketplace competition. In certain areas, the policy-makers believed, telcos were better positioned to develop new, cutting-edge technologies that would benefit consumers. As policy analyst Joel Rosenbloom explains:

> The FCC tentatively found . . . that phone-company entry was likely to produce services to the public of a kind that the cable industry had not provided. There were new possibilities of broadband services, stemming in large part from the growing use of optical fiber (with its much greater transmission capacity). The Commission focused, in this regard, not only on non-video services for which broadband capacity was needed, but on switched video services.[4] It found the latter services particularly intriguing. While cable systems were beginning to employ optical fiber, they were not configured to provide switching and cable operators had no experience in performing that function. (270–1)

The 1996 Telecommunications Act then further opened the door for competition among cable, DBS, and the telcos. One of the goals in framing the 1992 Cable Act had been to encourage the telcos to offer a service called "video dialtone," essentially a common carrier service that replicated with audiovisual messaging what the telcos had always provided with audio only. Notions that this service might deliver actual programming in addition to interpersonal communication – while unfulfilled – harked back to AT&T's efforts at toll broadcasting in the early 1920s (just prior to the rise of broadcast networks NBC and CBS), which many media historians see as the foundation of the commercially sponsored broadcasting system that is unique to the United States.

The 1996 Act replaced the video dialtone concept with something called "open video systems" (OVS), which hypothetically would be a hybrid platform – between a common carrier and a program service similar to traditional cable television service. In offering OVS service, telcos would be subject to somewhat stricter FCC regulations than those that applied to cable and DBS operators. Rosenbloom summarizes the provisions as follows:

> There were to be some requirements with a common carrier flavor. If demand for system capacity should exceed supply, the operator (including any affiliate) would not be allowed to select the program services filling more than one-third of that capacity. The OVS operator

would be forbidden to discriminate in its carriage of video programming providers (except as required to fulfill PEG access channel and broadcast must-carry obligations) and would be required to provide terms to program providers that (with the same exception) were "just and reasonable, and . . . not unjustly or unreasonably discriminatory." Similarly, the OVS operator would be forbidden to discriminate unreasonably in favor of itself or an affiliate with regard to any material or information provided to subscribers for the purposes of selecting programming, or to omit television broadcast stations or other "unaffiliated video programming services" carried on the system from any navigational device, guide, or menu. (371–2)

This provision sounds very similar to the FCC's late 1960s and early 1970s rules regarding local program origination and the provision of public access facilities by local cable systems. One only has to look at the eventual outcome of those efforts to understand why nothing of this sort has come to fruition with the telephone companies.

Meanwhile, the cable industry itself had begun to explore ways of using its lines (first coaxial cable and then fiber-optic cable) to carry telephone transmissions. Their lines, cable operators argued, functioned as much bigger "pipes" for carrying information than did the thin, twisted copper pair lines that were connected to phones in businesses and residences. First, cable operators went after the urban business market for phone service. In 1992, cable MSOs TCI, Cox, Time Warner, Continental Cablevision, and Comcast launched a joint venture called Teleport to compete head-to-head with the telcos in providing telephone service to businesses in major urban markets. Since MSOS at this stage had not yet consolidated metropolitan areas into hubs, large cities typically were served by multiple MSOs, each with separate marketing operations. A joint venture such as Teleport, isolated from other, more established components of their businesses, would allow MSOs to pool resources for this experimental new service. Teleport did not prove as successful as the cable operators had hoped, however, and five years later they sold the business to AT&T. AT&T for its part was pleased by the opportunity to offer a type of local phone service after the 1984 prohibition against using its phone lines for local service. The cable companies had hardly given up on offering phone service at this point, though; in fact they had more reasons than ever before to do so – as will be discussed below.

Cable and the Information Highway

The cable industry also was thinking very broadly about interactive services and multiple channels of information in the early 1990s, when its members began bandying about the phrase "500 channels" (the origin of which is usually attributed to John Malone). The 500 channels concept stood for a nebulous array of promised cable service – from niche networks to on-demand programming to services that would place orders for take-out food. It was not unlike the Blue Sky discourses in this respect. The concept also prefigured the Internet, which was already being developed within a very different sector of society. Since cable companies were not yet aware of the fledgling Internet's potential, they began making moves toward interactive services that relied on technologies developed by and for them. Cable's premature aspirations to large-scale interactivity were most apparent in Time-Warner's "Full Service Network" (FSN) experiment launched in Orlando, Florida in December 1994. And TCI hastened its own entry into this arena by attempting to merge with telco giant Bell Atlantic (a deal that would collapse on a grand scale a few years later).

As of the early 1990s, only a relative handful of personal computer users were aware of something called the Internet. At this stage, the World Wide Web and e-mail were in the realm of academic researchers and the scientific community. The arrival of user-friendly browser programs, first Mosaic and then Netscape and others in the mid-1990s, began to change this. Browsers introduced the graphical user interface (GUI) concept, making what Internet content was available (e.g., instructions on how to dissect a frog) easier to locate and use. What followed – at an astonishing pace – were more popular websites ranging from shopping to government documents to reference resources. Slowly, the cable industry came to realize that the 500-channel concept was being supplanted by something much, much bigger.

As websites grew more interesting and complex, however, they took longer and longer to transmit via the telephone modems that were available to home users. Even with a 56.6 kilobytes per second modem (the fastest available to a home user in the mid-1990s), an Internet user could wait as much as an hour or more for a single page to load – never mind moving through the links on that page. It seemed that the more interesting and captivating the page was, the longer it took to be viewable! There clearly was a market for much faster data transmission than traditional twisted pair phone lines would allow, and so cable's bigger "pipes" came back into the interactivity picture. This is how cable companies found the new service that would shore up their financial interests heading into the converged media scenario of the twenty-first century.

Continental Cablevision launched the first cable Internet service in 1994 on its Cambridge, Massachusetts system. Within the next few years, all the other large MSOs had followed, including a service called @Home that was started by TCI and adopted by seven other companies. Time Warner introduced the similar Road Runner service between 1997 and 1998.[5] Then, within the next five years, all of the larger MSOs would begin exploring VoIP (voice-over-Internet protocol) as a means of providing phone service in competition with the telcos. And by the mid-2000s, MSOs were flocking to a marketing practice nicknamed "überbundling," in which cable subscribers would not only pay a flat rate for multiple tiers of analog and digital television channels but also, presumably, save money by ordering cable, high-speed Internet, and telephone service all from the local cable company. Just as the 1996 Act helped telcos by offering them a means to enter the multichannel television programming arena, so too it offered cable companies the legal means to enter telephony.

Digital Cable and Direct-to-Home Satellite, Phase II

Another major breakthrough during the 1990s was the advent of *digitization*, whereby the same conversion of information into binary code (zeroes and ones) that allowed computers to process and transmit information became available for transmitting television programming. In analog transmission, the method used by virtually all broadcast and cable media prior to the 1990s, a sound or image could be reproduced on a television set only imperfectly, due to the inevitable loss of information by the transmitting technologies and the further loss due to the distance a signal had to travel to any and all receiving technologies (e.g., the broadcast affiliate station, the cable headend, or the home receiving dish, and ultimately the home television set). A digital signal, in contrast, involves no actual *retransmission*, but rather the decoding of a set of instructions on how to *re-create* the original information at the receiving end.

Not only did digitization thus bring the promise of a far clearer picture for the home television viewer, it also promised the various television industries the opportunity to deliver far more of this high-quality programming than they had been able to do with analog transmission. Analog transmission requires entirely new images to be sent at split-second intervals, whereas digital transmission only sends portions of the image that have actually changed. For example, in a television program, characters might move around the set, but the set itself (usually comprising the bigger portion of an image) remains unchanged for several seconds or even minutes. So with digital transmission, multiple signals can be carried simultaneously on a single bandwidth. As of 2007, broadcast television entities are still a long way from realizing the full potential of digital technologies. But cable and DTH satellite operators have made more progress, exploiting the fact that digitization allows many more signals to be carried on single satellite transponders in order to introduce popular digital tiers of service. Complementarily, programmers have been able to spin off new specialty channels, available only in the digital tiers, that would not have been possible previously.

The advances in digital television technology represented a mixed blessing for the cable industry. While offering a number of advantages, digital cablecasting has called for some expensive new equipment – most notably computer-equipped set-top boxes capable of converting a digital signal to analog so that it can be viewed on analog television sets (which are still used in the majority of US households as of 2007). But at the same time, digitization has allowed cable companies to add service tiers filled with the new digital networks and additional digital services such as "on-demand" pay-per-view, which allows viewers to select viewing time as well as to start and stop a program multiple times during a 24-hour period (a clear precursor to the streaming video that would be available over the Internet only a few years later).

Meanwhile the new direct broadcast satellite (DBS) industry was growing by leaps and bounds. As discussed in the previous chapter, up until approximately the mid-1990s, DTH had referred to the large C-band dishes used mostly by rural residents lacking access to a local

Figure 7.2 Ku-band (DBS) satellite dishes on the side of a house [author photo]

cable system, and who had the yard space to accommodate those dishes. Eventually these subscribers were paying for programming packages similar to what was offered to cable subscribers: somewhere around 24–48 channels (though lacking access to local broadcast signals without the use of a broadcast antenna in addition to the satellite dish). All of this changed with the rise of Ku-band satellite service, also known as DBS or direct broadcast satellite in the early 1990s.

Ku-band transmissions are much more powerful than C-band transmissions and thus require smaller and less expensive dishes (18–20 inches in size and costing around $200). For this reason, DBS had proved a better rival for cable than C-band DTH since a Ku-band receiving dish can be placed on the side of a house or apartment. DBS providers also tend to offer a wider range of programming packages than cable operators. The downsides of DBS include: upfront equipment costs, installation difficulties (including attaching the dish as well as orienting it properly), and susceptibility to signal fade with rainfall. Cable and DBS both have their relative advantages and disadvantages.

The general concept of *direct broadcast* satellite – with its own dedicated transmission frequencies, as opposed to the ad hoc use of headend-type C-band dishes by home users – dates back to a notion of direct-to-home satellite service using the C-band, around 1980, when the Satellite Television Corporation (STC) submitted a proposal to the FCC. The proposal was accepted two years later, and the FCC announced that it would begin accepting applications for the new type of service. Of the thirteen original applications, eight were approved (based on the number of available orbital slots at the time). However, by 1990, virtually all of the successful original DBS applicants (a group that included such major corporate entities as Western Union and RCA) had decided that costs would be too high and acquisition of programming too difficult to make the venture feasible. Meanwhile, entrepreneur Charlie Ergen, who had gone into business distributing C-band dishes to home consumers in 1980, had his 1987 DBS application approved and begun distributing cable-type programming packages for a monthly fee. Around the same time, News Corporation's Rupert Murdoch had launched his Sky DBS service (now BSkyB) in the United Kingdom. US cable operators were beginning to pay attention to the possibility of new competition.

A coalition of cable MSOs launched the DBS operation Primestar in 1991 as a hedge against the possible competition from other DBS operators. Like Echostar, Primestar used the C-band frequencies and, with a service almost identical to the operators' cable packages, drew subscribers only from among those rural populations not yet reached by cable systems. Competition for Primestar would emerge only after the 1992 Cable Act forced cable networks (who were financially integrated with cable operators in every possible way) to make their programming available to DBS operators on equal terms.

By 1996, Primestar had been joined by competitors DirecTV and United States Satellite Broadcasting (USSB), both launched by Hughes Aircraft, and Echostar (whose programming service is called DISH Network). This appears to be the maximum number of DBS providers the US market will support. Primestar folded in the late 1990s, and as of 2007, the US DBS market is controlled by DirecTV (which absorbed USSB) and Echostar/DISH. News Corp. acquired a controlling interest in DirecTV and its parent company Hughes Electronics in 2003, after lengthy negotiations with former owner General Motors. With DirecTV, it also gained control of PanAmSat, a major satellite distributor in the Americas. It was the fulfillment of a dream for Murdoch, according to journalist Wendy Goldman Rohm, who explains, "DirecTV was key to building his coveted global satellite platform. He had no presence in the United States with satellite, though his British Sky Broadcasting – known as BSkyB – combined with Asian broadcaster Star and Sky Latin America, had most of the rest of the world covered" (5).

Another DBS programming service, the Christian-oriented Sky Angel, is independently owned and operated but uses Echostar equipment. Currently all DBS broadcasts are in digital format, placing this medium ahead of both broadcast and cable television in at least one respect. As of 2005 about one in four multichannel television subscribers was a DBS subscriber.

The International Scene

It is important to keep in mind that the businesses, policies, and technologies discussed so far in this chapter describe a scenario that is fairly unique to the United States. Internationally, these new technologies have entered into some very different established media scenarios – and thus have been adopted for uses that in some cases are very similar to the United States, but in others are strikingly different. For one thing, it was not until the 1980s or later that most nations allowed private, commercial television entities to compete with established national broadcasters at all. So naturally television consumers in those nations would have held some different expectations for television service. One might even go so far as to say that the launch of private, commercial *broadcast* channels (longstanding entities in the United States and Canada) have been other nations' equivalent of the rise of satellite cable channels in the United States, since these were the first channels to enter into competition with the publicly funded channels.

Thus looking at multichannel television from an international perspective readily highlights what a uniquely North American phenomenon "cable television" really is. In the United States and Canada, a far-reaching wired infrastructure was well in place by the time it became feasible to distribute additional television channels via satellite. For these countries, direct broadcast satellite came about, primarily, as competition for cable and, secondarily, as a means to serve the most isolated rural households. Yet Canada has struggled to maintain a national broadcasting system in competition with both its own private channels and channels originating in the United States.

Things have played out differently still in Mexico, even though cable television has been available there since the mid-1960s. Unlike in the United States and Canada, DBS is overtaking cable in Mexico. As the existing coaxial cable plant grows more and more outdated, it is proving more feasible for television providers like Televisa to replace cable lines with the more broadly reaching DBS technology. Certainly more of the scattered rural population can be reached this way (see Sinclair: Ch. 2). Similarly in Brazil, the colossal private corporation TV Globo has been using satellites to distribute programming across the nation's huge land mass.

In some other nations, if coaxial cable networks existed at all, they only began to be used for US cable-type television around the same time that direct broadcast satellite became available. For example, Germany had been using coaxial cable both to extend the reach of its state and regional public television channels for decades, along with bringing in additional channels from bordering nations. Less than a decade after the arrival of private broadcast channels, more than three-quarters of German households were receiving these in packages, along with the preexisting public channels, via cable. Direct satellite services with additional channels have only recently started to draw significant numbers of subscribers there. Italy, on the other hand, had never been wired for cable to any significant extent (due to a combination of geography and politics), and thus adopted subscription satellite services very quickly once they became available in the mid-1990s. That media magnate Silvio Berlusconi, who heads the private Mediaset corporation, also was elected Prime Minister around this time was a major factor in Italy's rapid adoption of the latest multi-channel technologies.

Satellite-based media technologies have been credited not only with helping bring an end to communist governments in Eastern Europe around 1990; they also have played a role in the cultural and economic conversion of those nations to capitalism. There was no way to stop the influx of terrestrial and satellite-carried commercial channels from Western Europe and elsewhere. The weak economies of the newly capitalist nations of Eastern Europe could hardly generate programming of the same quality as the imported services (ranging from the BBC to MTV).

Still, western media entrepreneurs who had first perceived Eastern Europe as fertile ground for development of new businesses wound up stymied by the degree of newly formed nationalist sentiment that confronted them. Western media were not as easily or eagerly adopted as one might have predicted. There were entrenched media practices and infrastructures to be dislodged, after all, as well as newly articulated nationalist sentiments. As *New York Times* reporter Roger Cohen wrote in December 1992:

> Tensions have centered on television, a powerful political tool. Ambitious plans to end state television monopolies and set up private networks have stalled amid political vetoes and procrastination. This maneuvering reflects the fact that in societies still nervously coming to terms with pluralism, the degree of independence of news organizations remains a fiercely contested issue . . . [P]oliticians raised in a world where news organizations were vehicles of propaganda still tend to view them essentially as a means to peddle their views. (Cohen).

Cohen also makes it clear that not just news organizations were affected by the political turmoil; The Family Channel (still part of Pat Robertson's media empire at that point) was also rejected. Concerns about foreign ownership accompanied concerns about foreign ideologies.

In mainland China (PRC) as well, the trans-border flow of television signals has challenged the political regime. China has been more successful than the former Eastern bloc at upholding its form of government – though hardly without difficulties. Ironically, one way in which the Chinese government has tried to combat the reception of satellite signals via home receiving dishes has been to build local cable systems carrying both state and carefully screened commercial channels. Chinese officials have learned that preventing citizens from using home dishes means providing equal or better quality programming via the cable systems – so in this case competition has had an indirect, though nonetheless palpable effect on program selection.

Not surprisingly, the cost of improving the quality of state-supported programming has led to increased commercialism in Chinese television overall (Thomas: 171–4). This, of course, has gone hand in hand with China's growing trade relations with the West in recent years. It might not be long before Chinese television is indistinguishable from that of fully capitalist nations – in terms of increased advertiser support, diminished government oversight, and more entertainment-driven program genres.

In parts of the world where television was even less developed as of the 1990s, the arrival of direct satellite services – typically in the form of regional services such as Murdoch's pan-Asian Star-TV – have proven jarring to established cultural norms and policies. Entities existing specifically to offer television across borders represent a new and formidable phenomenon that twenty-first-century citizens and governments will need to grapple with more and more. Their operations have evoked concerns reminiscent of those brought on by nation-to-nation importing and exporting of media products.

Global Television

Historically, transnational television entities developed for a variety of reasons. In some cases, transnational broadcasting has been a longstanding function of existing networks. For example, the British Broadcasting Corporation's international operations date back to the early days of radio. If Great Britain has lost political control of most of the nations it once colonized, the familiar phrase, "the sun never sets on the British Empire," still holds true of the cultural ties maintained (or in some cases initiated) through the BBC World Service. Particularly well known for its newscasts, it addresses regional audiences worldwide. The Voice of America (VOA) from the United States has similar overseas operations, though some see the VOA as maintaining a stronger political mission than that of the BBC.

In other cases, especially more recently, transnational television entities have resulted from the expansion of established commercial broadcast or satellite networks into new markets. For instance, the 1990s saw the conversion of US cable networks such as CNN, HBO (and other Time Warner premium channels), and Discovery into "global" channels once they had reached maturity in the domestic market.

CNN established its reputation as an up-to-date international news source during the 1990 Gulf War, and two years later spun off CNN International as an English-language service with five discrete regional feeds: Europe/Middle East/Africa, Asia Pacific, South Asia, Latin America, and North America. International versions of other US-based satellite networks maintain varying proportions of dubbed or subtitled US-produced content, which tends to be quite popular internationally, and content oriented toward the specific countries being served. The balance can be related to the production resources available locally. HBO Brazil, for example, offers a Brazilian-produced series called *Filhos Do Carnaval* in addition to imported movies and series, and HBO Argentina ran the locally produced series *Epitafios* in 2005. Other national versions of HBO offer only imported programming – both from the United States as well as from other national exporters. Both *Filhos do Carnaval* and *Epitafios*, for example, have been shown on HBO networks throughout Latin America and elsewhere.

Some international networks have come into existence in the recent past specifically to serve audiences in more than one nation. The ascendancy of the European Union (EU), especially as it draws in more nations from the former Eastern bloc, has tempted large media corporations to target audiences across national borders in order to increase scale economies. As Thomas L. McPhail explains:

> These combined and larger media companies are in a better competitive position because they can offer either larger audiences to advertisers, or a larger number of cable subscribers to generate revenues necessary to upgrade cable systems so that they are internet-ready . . . [M]ore commercial corporations are designing advertising and programs for a pan-European audience. Advertisers want to deal with major trans-European broadcasters for a single package rather than with small individual media outlets on a city-by-city or country-by-country basis. (99)

Among the most prominent European border-crossing media conglomerates are Bertelsmann/RTL Group (Germany), Vivendi/Canal+ (France), and Fininvest/Mediaset (Italy – the company formerly owned by Silvio Berlusconi). These conglomerates all are in the business of producing and distributing television programming both in Europe and worldwide.

If distribution technologies, along with program content tailored to those technologies, have been a driving factor in the spread of transnational networks, Rupert Murdoch and News Corporation have been at the controversial center of this trend. As discussed above, Murdoch launched BSkyB in 1989 as a means to distribute an array of private, commercial television channels in the United Kingdom. Shortly thereafter, in 1993, News Corporation acquired a controlling interest in the pan-Asian Star-TV satellite service. Based in Hong Kong, Star had been started three years earlier for the purpose of distributing national and transnational networks broadly across Asia. Today several of its channels reach into the United States, Europe, and the Middle East as well – providing Asian-themed programs in English as well as a variety of Asian languages.

Complementing the role of these satellite distributors are the global programmers themselves. The notion of dedicated radio and television *stations* or *networks* to serve specific language or cultural communities within a nation (known as "diasporas" in the case of minority populations) is not a new one; in fact it goes back to the early days of radio. However, it was economically unfeasible to operate foreign-language broadcast stations where there was no critical audience mass to sustain them. So where Spanish speakers in cities like San Antonio or New York had programming available through such entities as the Spanish International Network (SIN), those in cities like Dubuque, Iowa or Charlottesville, Virginia did not. The same was true of diasporic communities worldwide. DBS has changed this situation on an international scale. With satellite footprints covering large land masses, language- or culture-specific channels can be made available to anyone with a satellite receiving dish. Many travelers in the United States have been surprised to find Hindi-language programming among the channels available at rural roadside motels!

While English-language programming remains predominant in the export markets, it clearly has competition from programming produced in other widely spoken languages. By the 1990s, Latin American television giants like Televisa and Globo were well positioned both to distribute a range of foreign-originated channels throughout their home nations and also to export programming to Spanish- and Portuguese-language populations throughout the world. The pan-Arab network Al Jazeera (see Box 7.2) now serves audiences in Arab-majority nations in the Middle East, Asia, and Africa, along with

Box 7.1   Al Jazeera

If the 1990 Gulf War cemented the notion of CNN as an authoritative source of international news, the Middle East-based Al Jazeera satellite network underwent a similar transformation during the wars of the late 1990s and early twenty-first century. Al Jazeera (a name meaning "the island" in Arabic) has confidently challenged the notion that a US- or western-centric version of events is the only one to be known – actually resulting in some westerners accusing the network of being "the mouthpiece of Osama bin Laden" (a misunderstanding that has diminished somewhat). So thorough and revealing has Al Jazeera's live war coverage been, in fact, that other news entities, including CNN as well as the BBC and other Western networks, have used its footage in their own reporting.

Al Jazeera was founded in the small Persian Gulf nation of Qatar in 1996. This was the year in which the Emir Sheikh Hamad bin Khalifa Al Thani seized power from his long-reigning father, Sheikh Khalifa bin Hamad Al Thani, and subsequently initiated a series of political reforms that, while not dislodging the conservative forms of Islam that prevailed in the country, made some modifications to the archaic and strict codes of both social and political behavior in place at the time. Al Jazeera was part of these reforms. It was launched with a $137 million grant from the Emir, and while treated as an independent journalistic entity, continues to be subsidized by the Qatari government. Many of its founding staff came with experience in international journalism from working for the BBC World Service, known for extensive and balanced coverage of international news. The goal for Al Jazeera has been to uphold a similar set of values and standards – but producing this sort of journalism from within the pan-Arab culture itself.

Throughout its short history, Al Jazeera has strived to maintain a position of neutrality in its coverage of various conflicts in the Middle East. Its advertising and marketing arm explains: "Al Jazeera offers their viewers a different and new perspective, it was the first of the Arab TV stations to break the unwritten rule that one does not criticize another Arab regime, the source of much of its earliest controversy. But regardless of what their rulers thought, viewers were delighted to get something other than the usual pro-government propaganda" (History of Al Jazeera Television. Allied Media Corp. http://www.allied-media.com/aljazeera/JAZhistory.html).

For example, in 2000, at the start of the second Israeli intifada, Al Jazeera initiated the unprecedented practice among Arab media of interviewing both Israelis and Palestinians on issues related to the conflict. Journalist Hugh Miles remarks: "One key difference between Al Jazeera and the other Arab channels was its policy of interviewing Israelis. It was still the only Arab channel to do this [as of 2000]. Other Arab channels still either spoke for the Israelis or ignored them completely. This policy drew severe criticism from the Arab world . . . [But] Al Jazeera was defiant, saying that, in keeping with its motto, the Israelis had to be given a voice" (92– 3). On some occasions, Al Jazeera has been banned from certain nations, at least temporarily, because of its unwillingness to take sides. And some Al Jazeera journalists live in fear of violent retribution for allowing controversial political and religious figures to express their views on the network.

Al Jazeera has gained much notoriety with its coverage of the Afghanistan and Iraq wars of the early twenty-first century. Nevertheless, many North Americans and Europeans have been unaware of the extent to which Al Jazeera reporting has affected their understanding of events in the Middle East. Perhaps this has begun to change following the November 15, 2006 launch of Al Jazeera's English-language international service, Al Jazeera English. Al Jazeera also maintains an English-language website: www.aljazeera.net/english.

minority Arab populations elsewhere. A somewhat similar function is served in the region known as "Greater China" by Hong Kong-based Phoenix Satellite TV (partly owned by News Corp.). Phoenix presents itself as a regional alternative to Chinese Central Television (CCTV), mainland China's state broadcaster, and its satellite feeds cover virtually the whole world. In addition to its proprietary channels, Phoenix distributes an array of other Chinese-language (Mandarin and Cantonese) services.

That the desire to tap into international markets is true of most media corporations operating today seems an appropriate observation with which to end this chapter – which does, of course, cover a time

Figure 7.3 Al Jazeera International's Doha, Qatar studios [Al Jazeera International]

period that is ongoing. The long-term implications of media convergence and globalization can only be speculated upon. And since this sort of speculation seems the only appropriate "conclusion" for a book of this type, that will be the goal of the next and final chapter.

NOTES

1 Note, however, that in summer 2006, Time Warner stock dropped upon the company's announcement that more and more of what had always been AOL subscription-based online services would need to be offered for "free" (subsidized by advertising) in order for the company to remain competitive with such giants as Yahoo and Google.
2 For more details on these mergers and acquisitions, see Croteau and Hoynes: Ch. 3.
3 United States Telecom Association: http://www.ustelecom.org/index.php?urh=home.news.telecom_stats.
4 With both broadcast and cable television, all channels are delivered to the home, and the viewer chooses which to watch by "tuning it in." With "switched video," however, the telephone company as video provider sends only one

channel at a time to the viewer. The viewer essentially makes a request to have one particular "channel" of streamed digital programming sent at one time.

5   While Time Warner continues to provide Road Runner service, other MSOs have since launched their own proprietary high-speed Internet services.

FURTHER READING

Aufderheide, Patricia. *Communications Policy and the Public Interest: The Tele-communications Act of 1996.* New York: Guilford, 1999.

Croteau, David and William Hoynes. *The Business of Media: Corporate Media and the Public Interest.* Boston: Pine Forge, 2001.

Sinclair, John and Graeme Turner. *Contemporary World Television.* London: BFI, 2004.

Thussu, Daya Kishan. *International Communication: Continuity and Change*, 2nd ed. London: Hodder Arnold, 2006.

Chapter 8

The View from 2007: The Future of Multichannel Television

Looking back over multichannel television's nearly six-decade history in the United States and elsewhere, it is clear that this is the story of a "making do" industry evolving into a "can't do without" industry. Where rudimentary community antennas first appeared throughout North America as a means of making television channels available in places where they would not have been otherwise, today's cable and satellite television industries encompass an array of entertainment and information sources and services not available through any other means. Where small-town CATV operators once feared that a policy decision would shut down their industry entirely, today the cable industry (so tightly integrated with other media and telecommunications industries) represents a powerful lobbying force in Washington, DC and other seats of government throughout the world. And where CATV service once was known almost exclusively to residents of rural areas, today cable or satellite service (and all that implies) is considered indispensable to many people in urban areas as well.

Unlike broadcast television, which was developed primarily in the laboratories of established radio corporations, cable television was developed in the homes or shops of tinkerers, most of them electronics amateurs. Information was shared through a remarkably organized and cohesive trade organization, the NCTA. On its way to becoming cable television (and sharing its industrial and programming precedents with other forms of multichannel television), CATV traversed some extremely mercurial policy terrain. At every stage, operators and others affiliated with the industry found creative solutions to working within regulatory constraints. The conceptual shift from

"CATV" to "cable television" was helped along by a shift in the pre-dominant US policy-making climate from left to right. Cable's leap into the satellite era represented a combination of factors, including the Nixon administration's deregulation of both cable and domestic communications satellites, the presence of programming operations such as HBO and the TBS superstation that were already in place and ready for uplinking, and advances in the technology itself – both satellite and cable.

Cable's next leap would be into the era of integrated media opera-tions. The cable industry was gathering a great deal of strength at the same time that huge mergers within the media industry writ large were beginning to occur. For example, cable operations rapidly were absorbed into the accruing synergy of the newly merged Time-Warner conglomerate in 1989. And cable proved itself an ideal player in this scenario, functioning as both a distributor and exhibitor of media products – at first movies and television programs, and eventually music, video games, and even the Internet itself. The infrastructure and technologies that allow cable and other multichannel forms of television to function today only enhance the integration of the various media. An enhanced cable subscription today can provide or replicate the function of virtually every other form of electronic media used in a modern household, from video rentals to phone service to libraries to shopping. Moreover, in the near future, it is likely that cable itself will exist as a concept only, and probably an archaic one at that. Already most industry professionals prefer the term "broadband" to describe the array of information services offered through cable, DBS, and now telephony.[1]

So imposing an end to a book about the history of multichannel television is a daunting challenge, since the twenty-first century in many ways represents much more of a beginning than a conclusion to an era of multichannel possibilities. As of 2007, it looks as though the cable/satellite era is about to end – just as the CATV era did in the 1970s. Throughout the later part of the twentieth century, the term "multichannel" generally referred to a single medium or mode of transmission – broadcast, cable, satellite – having the capacity to offer viewers more than a single channel of content. The more channels the better, and hence the desirability of cable and eventually direct broadcast satellite over broadcast television. In the twenty-first century, however, it is already clear that "multichannel" is giving way to "multimedia" or "multiplatform" as the way to characterize the most desirable media scenarios for consumers.

Broadband Media in the United States in the Early Twenty-First Century

As of 2007, cable and direct-to-home satellite operations still flourish (there are even a handful of C-band satellite dish users still around). But increasingly the companies that provide these media see their future successes as inevitably linked to an array of other media technologies such as the Internet, telephony (wired and wireless), and numerous other electronic media. Already there are indications that the future marketplace viability of media content will lie in the flexibility with which that content can be accessed – perhaps even more than in the nature of the content itself. Television audiences "cast our vote" for flexibility when we began using videocassette recorders for time-shifting and commercial "zapping" during the 1980s. The prerecorded television shows, digital video recorders, and on-demand pay-per-view programming that have emerged since then only reinforce the point: Consumers desire media products as lifestyle enhancements more than as lifestyle determinants. Perhaps we always have wished for this flexibility, but simply lacked the technological capability for it in the past. Very few people nowadays set their personal and family timetables based on which television shows they want to watch – as was common practice in earlier decades of television history. More and more people access their preferred forms of electronic entertainment in nontraditional spaces (public transit, doctors' waiting rooms, etc.) using such devices as portable DVD players (or DVD-capable laptop computers), portable gaming devices, iPods and MP3 players, and even cell phones and PDAs. And more and more public spaces such as supermarkets, cafés, and airports provide means of accessing electronic media content – ranging from publicly viewable large-screen televisions to WiFi hotspots.

The traditional home audience is no less affected by changes in the ways media products can be accessed. Even as television sets and their accompanying cable or satellite boxes grow more similar to home computers in the technologies they encompass, more people are turning to the home computers themselves as places to access audiovisual information and entertainment products. For years it has been possible – often desirable – to watch DVD movies on computers instead of television sets. Now television-type content is available (using streaming software) directly through the Internet. The timeframe within which this phenomenon is evolving is surely one of the shortest ever witnessed for a television technology. At the start of 2005, streaming video was cumbersome for most Internet users, even those with high-speed connections. It was useful for short clips or representative samples of video products, but was hardly a replacement for television. A year later, websites such as Google Video and Nick.com began to offer full-length television programs that could be viewed (in small format) using software programs such as Windows Media Player, Quicktime, and iTunes. In 2007, a much broader array of programming is available via the Internet – to those willing to put up with the still somewhat clumsy technology – than through any other video medium.

Figure 8.1 More and more people are choosing to watch the range of video programs available via the Internet, either in place of or in addition to broadcast and satellite television channels [Getty Images]

The YouTube phenomenon arose almost overnight during late 2006. Through the YouTube website (launched in February 2005), users may upload virtually any short videos, ranging from video-camera practice footage to pirated clips from "regular" television to clever and inspired efforts by aspiring film and video producers. YouTube no doubt has a strong appeal to the short attention spans that characterize members of the "Millennial" generation (today's teens and young adults), a cohort already two generations removed from a television audience not socially predisposed to channel-surfing. Easily the most talked about media business deal of 2006 was the phenomenal Internet search company Google's October announcement that it would purchase YouTube for $1.65 billion in stock. Many analysts noted ways in which this merger would integrate broadband-type video services with other spheres of the online information and social environment. *CNN Money*'s Paul R. LaMonica reported at the time of the announced acquisition that:

> The combination of Google and YouTube could further strengthen Google's dominance in online advertising, giving it an edge over rivals such as Yahoo!, Microsoft's MSN and News Corp., which owns the social networking site MySpace. Some analysts said Monday that Yahoo, Microsoft and News Corp. also had probably expressed interest in buying YouTube.
>
> Bill Tancer, general manger of global research of Hitwise [a market research firm], said after the deal was announced that MySpace would now probably need to promote its own video service more aggressively on its site in order to compete with a combined Google-YouTube. (LaMonica 2006a)

Interestingly, LaMonica had already noted in a July 24, 2006 article that would-be competition for YouTube was emerging in the form of outfits such as Gotuit Media that were interested in specializing in professionally produced online video content:

> Along those lines, Gotuit.com will debut with music videos from major record labels such as Geffen, Interscope and A&M, news from Reuters and the Associated Press, and movie trailers from Sony (Charts), News Corp.'s (Charts) Fox and Warner Bros., which is also owned by Time Warner. The service is free to use and the company hopes to make money from online advertising.

But [vice president of product management Patrick] Donovan said that what he thinks will really separate Gotuit from companies like YouTube is its technology. The site indexes all its videos so that people can easily start watching a clip of a video at a specific point.

Videos are also laid out in a way so that viewers can click on a new one while watching another and have it instantly start. In that sense, it's more like using a remote control to watch television instead of sitting around waiting for a video stream to buffer.

"This is premium content coupled with a premium viewing experience," said Donovan. "It combines the best of TV with the control of the Web." (LaMonica 2006b)

Clearly what computer users and television viewers are witnessing in the early twenty-first century is a negotiation of how much and what sorts of video content will be marketable, as well as how that content will be delivered most effectively.

Cable companies for their part are ready and eager to deliver an integrated range of information, entertainment, and telecommunications services. As of early 2007, they hold a slight lead over their telephone company competition with their "überbundling" strategies. Now on the horizon is the addition of wireless phone service to this sort of package. And Telcos are rapidly closing the gap. Though still technically prohibited from offering cable, they now offer strikingly comparable services. For example, AT&T's U-verse and Verizon's FiOS services both claim to integrate seamlessly cable-type digital video services and computer gaming with broadband Internet. On its promotional website, Verizon likens the innovative qualities of FiOS to inventions ranging from fire to the airplane to computers. Time alone will tell if these services offer true competition for cable MSOs, since U-verse and FiOS are still available only in limited test-market areas.

Along with the new incarnations of television distribution, it is important to consider how televisual media *content* has changed in tandem with changes in viewing habits. For one thing, the various forms of multichannel or broadband video have reached a level of full competition with traditional broadcast television, chiseling away at the older medium's market dominance. Broadcast television thus has had to grapple with the issue of its own relevance and ability to compete. Even a decade ago it would have been almost unthinkable for a broadcast network news personality to resign and move to a cable/satellite network. Given the presence of successful online video providers, it is even worth questioning whether cable networks themselves are outdated. But this is precisely what *Nightline*'s Ted Koppel did when he left ABC at the end of 2005. See Box 8.1 for Koppel's reasons for moving to Discovery Networks, as explained to journalist Ken Auletta in January 2006.

Box 8.1   Interview: Ken Auletta discusses Koppel moving to Discovery Channel, "Talk of the Nation," National Public Radio, January 4, 2006

NEAL CONAN, host:
Speculation about where newsman Ted Koppel would land once he left ABC's "Nightline" ended today with the announcement that he's joined the Discovery Channel. Koppel was named managing editor of Discovery, and he will produce documentaries for the cable channel. Joining us now to talk more about this is Ken Auletta. He writes the Annals of Communications column for The New Yorker magazine. He's the author of several books about the news media. He's with us on the phone from New York City.

Ken, thanks for taking the time to speak with us today.
Mr. KEN AULETTA (The New Yorker): My pleasure.
CONAN: Are you surprised that he ended up at Discovery? There were a lot of rumors about HBO.
Mr. AULETTA: Yeah, the rumors were HBO and so it is a surprise. But my understanding is that Discovery came in relatively late in the bidding process with a more substantial offer and more freedom.
CONAN: Discovery is not a place that's exactly a beacon of journalism.
Mr. AULETTA: It is not. I mean, it does wonderful documentaries, most of which they farm out and basically pay for, consultants, in effect, and oftentimes, you know, life – you know, environmental documentaries, etc. Koppel is – and his staff of eight, including Tom Bettag, his executive producer, is actually going to be housed in the headquarters in Washington of Discovery

Network. They have a fairly big audience as does, obviously, HBO for cable networks, and they're obviously choosing to go in a different direction and expand their – the stuff they do that they actually own.

CONAN: That's interesting 'cause, as you say, Discovery farms out almost everything. I used to think that their corporate goal was to have no one eligible for health insurance.

Mr. AULETTA: (Laughs) Yeah, it is a form of outsourcing, isn't it?

CONAN: Yeah, yeah. We're talking about Discovery Channel and its acquisition, I guess, of a news department, a managing editor, Ted Koppel, today.

You're listening to TALK OF THE NATION from NPR News. And it's interesting, Ken Auletta, when broadcasters tried to develop news departments in the past, what they did was develop a news department – ABC News, CBS News, NBC News – I guess – modeled, I guess, on the old newspaper, but this is a different model.

Mr. AULETTA: This is a different model and it's actually one that – there's an undercurrent here that's real important, I think, Neal, to make clear. Koppel has basically – is implicitly saying that the world he comes from, the world of network broadcast television news, is not interested any longer in the kind of serious news that he wants to do. And it is – they are places increasingly that are preoccupied with their audience, which is dwindling, and reaching a younger demographic. Now the kind of international documentaries or "Nightline"-type interviews that Kop – or town halls that Koppel likes to do are not ones that the networks are very interested in doing anymore. Discovery is saying, "We are interested in doing it." So Koppel is really saying, "In the world of network television that I grew up in and succeeded in and was paid a very healthy amount of money in, that world is no longer hospitable to Ted Koppel and the kind of news I want to do. But Discovery and the cable networks, which can charge and have two sources of revenue, not just advertising but obviously subscriptions of some kind, that that's the world that I might be able to thrive in."

CONAN: You can see what's in it for Koppel. What's in it for Discovery?

Mr. AULETTA: Well, they – you know, what's in it for HBO when they put on great programming like "The Sopranos"? They get – you see, people talk – in business talk about brands all the time. But what does a brand really mean? In the news world, a brand means credibility. Ted Koppel gives instant credibility to Discovery. So if Discovery is saying, 'We want to do our own news,' what better brand or source of credibility than Ted Koppel?

CONAN: So it's a prestige buy?

Mr. AULETTA: It's – well, it's more than a prestige buy. It's also a potential audience lure. Don't forget, they don't have to achieve in the cable world the mass audience. And everyone talks about FOX News and how well it's doing vs. CNN, but, at most, they have a couple of million people watch at night. That's compared to a network – the network in the news division has 26 million people at the three – watching the three newscasts at 6:30 at night.

CONAN: And they're considered failures–or failing.

Mr. AULETTA: And they're considered failures. So they have a mass audience. In cable, you don't have to achieve a mass audience. So Koppel will enlarge the current audience that Discovery has, will bring them showers of publicity and credibility presumably and that, in the long run, might help Discovery create a new brand, a new awareness.

CONAN: And is this another step in the process – the long process but – whereby Channel 7 is no more significant than Channel 356?

Mr. AULETTA: Yes, that's right. Now the question I have about the Discovery agreement: Television, you know, where is the future component, where is the Internet component to this deal? How is Ted Koppel – does Ted Koppel and/or Discovery have any idea of how to get themselves not just on television or not – but on devices, on – via cell phone or a PDA or some kind of handheld portable device or a game console? That's where increasingly you see the world moving is people have more and more portability with their devices.

CONAN: Yeah, iPods . . .

Mr. AULETTA: Correct.

CONAN: . . . just in the last six months have changed. I mean . . .

Mr. AULETTA: Totally.

CONAN: Yeah, this – podcasting didn't exist two years ago.

Mr. AULETTA: It didn't. So the question is: Is it possible that we will see Ted Koppel in some kind of video form on our cell phone?

CONAN: A question to ponder. Ken Auletta, thanks very much for being with us today.

Mr. AULETTA: My pleasure, Neal.

CONAN: Ken Auletta writes the Annals of Communications for The New Yorker magazine. He's also the author of "Backstory: Inside the Business of News."

This is TALK OF THE NATION from NPR News. I'm Neal Conan in Washington.

Koppel's decision was probably a wise one for the short term, since middle-aged viewers, while generally willing to opt for a cable/satellite network over one of the major broadcast networks, might not be ready to switch from the television to the computer as their first viewing option. Broadcast networks have come to terms with the fact that they must compete as niche services among an array of other niche services. They increasingly are resorting to less expensive genres such as reality programs and talent contests – that have proven successful with a range of audiences – to stay competitive in serving a general audience. And, contrastingly, to the extent that the broadcast networks are still producing original dramatic programming, these tend to be very high-budget shows such as the *CSI* shows and *House* that are geared toward both the tastes of upscale audiences and the demands of the retail video aftermarkets, where these programs compete as much with feature-length movies as they do with other TV series. Interestingly, in the 1980s and 1990s, cable networks such as HBO, Showtime, TNT, and Hallmark Channel began using these same strategies as a way of branding themselves as producers of quality products as they courted their small slice of the television audience. Could it be that, in the rapidly shifting post-2000 video landscape, cable networks are already facing the same challenges their broadcast predecessors began to see just a couple of decades earlier?

The 2006 Telecommunications Bill

Clearly the US information and entertainment scenario continues to change – and government policy-makers continue to seek ways to control. Ten years after the passage of the 1996 Telecommunications Act, Congress once again began debating legislation to address the current status of telecommunications in the United States. Formally called the Communications Opportunity, Promotion and Enhancement Act of 2006 (COPE), this proposed legislation addresses the following key areas:

- National cable franchising: provisions that would establish a franchising process and set of standards applicable to every community in the United States. This would replace the locally determined franchising processes now in place, which the cable companies find cumbersome.
- Network neutrality: provisions to ensure that consumers have access to all Internet content, without preference being given to wealthy corporations as content providers. This provision would prevent Internet access providers from offering "premium" or "fast-lane" access to customers willing to pay more.
- e911 and VoIP (Voice over Internet Protocol): ensures access to emergency (911) service for subscribers to broadband Internet phone services.
- Municipal broadband: a provision to protect the rights of municipal governments to provide broadband Internet service.

The legislation was introduced in the House of Representatives on May 1, 2006, and was passed by that body on June 8. However, it died in the Senate in September of the same year. It is worth mentioning in this final chapter because it is clear evidence of how much the traditional policy-making climate in the United States continues to be stymied by the arrival of new media technologies – not to mention the increasing speed with which those technologies alter existing cultural practices and assumptions.

New media technologies have introduced concerns that cannot be addressed effectively by the US legislature – even though they involve US media corporations. These are concerns involving an increasingly globalized media scenario, in which any protections secured by and for US residents might not even be enforceable. Even if net neutrality provisions had been enacted, for example, how could this effectively safeguard consumers against any anticompetitive practices carried out by Internet providers based in other countries? Already online pornographers and gambling operations have moved their operations abroad – keeping them available to US Internet users but out of the jurisdiction of US enforcement agencies.

The Global Scene

If US residents have reason to be concerned about electronic media products originating in other countries, surely residents of other countries have more concerns than ever before about the media incursions that affect them. By 2007, it is clear that the multichannel entertainment and information arena of the future will be one in which national boundaries are increasingly irrelevant to how business is conducted. This hardly eradicates reasons for concern over cultural imperialism, of course, for information flows remain very uneven. The eroding of local cultures brought about by the importation of media products is troubling indeed. But whether or not transnational media operations *deliberately* erode cultural traditions is debatable, since border crossing serves an important economic function: the more consumers who can be reached by (and therefore help subsidize the production cost of) a media product, the more a producer can afford to risk on producing that product. Some would argue also that border crossing is a necessary precondition for the economic success of those corporations serving smaller countries or regions that lack local funds or significant economies of scale. Left to their own resources, those countries might produce only inferior quality television, if any at all.

When considering the cultural imperialism issue, perhaps one cause for optimism is that new program producers and distributors not based in the United States or Western Europe have emerged as major players in the global television market, especially as satellites allow programming to be distributed uniformly and inexpensively over large national and regional land masses. Regional lingua franca languages other than English – such as Chinese (Mandarin and Cantonese), Portuguese, and Spanish – have grown increasingly dominant in global television. One need only consider the dramatic expansion of entities such as Brazil's Globo, Mexico's Televisa, and Hong Kong's Phoenix Television during the 1990s to see evidence of this.

Even the many US-based satellite programming outfits that have expanded their operations overseas increasingly see a need to adapt to expectations at the local and regional levels. For example, while dubbed or subtitled US (or other English-language) movies and series still dominate HBO's many international services, there have been efforts to integrate series and movies with more connection to the part of the world being served. The series *Filhos do Carnaval* (Brazil) and *Epitafios* (Argentina) discussed in Chapter 7 are examples. These series typically are shown in their countries of origination as well as elsewhere in the region or linguistic community (*Epitafios* was shown in the United States on HBO Latino).

CNN International provides its five regional feeds which, while hardly catering to the news needs of specific nations, at least acknowledge that Americans' news preferences (or those of other Western nations) do not necessarily correspond with those of the rest of the world. Perhaps as more competition emerges, CNN International will become even more region- or nation-focused in its coverage. We might know very soon, since as of 2006, the new Qatar-based Al Jazeera International news service stands ready to compete with CNN International.

And even within the United States itself (as well as in other Western nations) inroads are being made by programming services for cultural and linguistic diaspora communities. So even as new media technologies continue to allow Western television entities to introduce foreign values into the cultures of other nations, those same technologies are helping to maintain the cultural identities of groups living abroad. The same technologies also are making the cultural traditions and program genres of other countries more available to those steeped in US culture – evident in the growing interest in international soccer's World Cup, the various genres of *anime* and *manga* (that originated in Japan), program imports such as *Iron Chef* (Japan) and *Trailer Park Boys* (Canada), and a host of other foreign programming that is available on mainstream US cable and satellite networks.

Time alone will tell what the concluding chapter to a future edition of this book might include. It no doubt would be similar to the present chapter in raising questions more than providing answers, for change is the one constant in media development. It is clear that multichannel television has grown into far more than the CATV pioneers of the 1950s could have imagined, as they tinkered with technologies that would capture distant broadcast signals and deliver those signals to homes with no other way to receive television programming. If the US cable television industry came into existence as a way to make sure America's rural residents had television service that was at least as good as that of urban residents, it and its conceptual heirs have continued to develop and deliver entertainment and information services are indispensable to most people. Many US households now pay a single monthly bill to their cable company that covers television programming, high-speed Internet service, and phone service. Surely once phone companies become full-fledged entrants into this market, the number of services will expand far beyond this.

Moreover, the US model of multichannel television that traces its origins back to the CATV pioneers has come to define multichannel television for virtually the whole world. Most nations have had access to the fundamental technologies of multichannel television (coaxial cable, for example) for decades. Some of those nations have even enlisted the technologies for delivery of television programming. However, it has only been in the past couple of decades that a more or less global set of expectations has emerged: that multichannel television is desirable, that it should be privately controlled and commercially supported, that it will cross national borders, and that it will offer more services with each passing year. The CATV pioneers who have lived long enough to witness this outcome merely shake their heads in amazement.

NOTE

1   The Cable Center defines broadband as "a transmission medium that allows transmission of voice, data, and video simultaneously at rates of 1.544Mbps or higher. Broadband transmission media generally can carry multiple channels – each at a different frequency or specific time slot." The Cable Center website. Cable and Telecommunications Glossary. Available from: http://cablecenter.org/education/library/glossary [cited January 6, 2007].

Bibliography

Adler, R. and Baer, W. S., eds., 1974. *The Electronic Box Office: Humanities and the Arts on Cable.* New York: Praeger.

Aufderheide, P., 1999. *Communications Policy and the Public Interest: The Telecommunications Act of 1996.* New York: Guilford.

Auletta, K., 2004. *Media Man: Ted Turner's Improbable Empire.* New York: Norton.

Baldwin, T. and McVoy, D. S., 1988. *Cable Communication,* 2nd ed. Englewood Cliffs, NJ: Prentice-Hall.

Barnouw, E., 1990. *Tube of Plenty: The Evolution of American Television.* New York: Oxford University Press.

Bibb, P., 1993. *It Ain't As Easy As It Looks: Ted Turner's Amazing Story.* New York: Crown.

Big-time venture in small-time TV, 1970. *Broadcasting,* November 9, p. 50.

Bruck, C., 1990. The world of business: Deal of the year. *New Yorker,* January 8, pp. 66–89.

Cable Television Laboratories, Inc., 1998. *A Decade of Innovation: The History of CableLabs, 1988–1998.* [Online] Available from: http://www.cablelabs.com/downloads/pubs/history.pdf. [Cited January 12, 2007]

Camire, D., 1992. Congress approves cable bill; Hands Bush a political slap. Gannet News Service, October 5. [Cited in Robichaux, p. 119]

Cantor, L., 1971. Pennsylvania system gets lots of volunteer help, *TV Communications.* July, pp. 94–100.

Chenoweth, N., 2001. *Rupert Murdoch: The Untold Story of the World's Greatest Media Wizard.* New York: Crown Business.

Cohen, R., 1992. Eastern Europe meets obstacles to free expression. *The New York Times,* December 27, pp. 1+.

Community TV group forms national lines, 1952. *Broadcasting-Telecasting,* June 16, p. 84.

Compaine, B. and Gomery, D., 2000. *Who Owns the Media? Competition and Consolidation in the Mass Media Industry,* 3rd ed. Mahwah, NJ: Lawrence Erlbaum.

Croteau, D. and Hoynes, W., 2001 *The Business of Media: Corporate Media and the Public Interest*. Boston, MA: Pine Forge.

Daniels, B., 1986. Oral history interview transcript. Interviewed by M. Paglin. Denver, CO: The Cable Center. Available from: http://www.cablecenter.org/education/library/oralHistoryDetails.cfm?id=218. [Cited January 12, 2007]

Dempsey, J., 1998. TW and TCI swap cable systems: Cluster deal affects 1 million subs. *Variety*.com, December 11. Available from: http://www.variety.com/article/VR1117489330.html?categoryid=18&cs=1. [Cited January 12, 2007]

Denisoff, R. S., 1988. *Inside MTV*. New Brunswick, NJ: Transaction Press.

Easton, K., 2000. *Building an Industry: A History of Cable Television and its Development in Canada*. East Lawrencetown, Nova Scotia: Pottersfield.

El-Nawawy, M. and Iskandar, A. 2003 *Al-Jazeera: The Story of the Network That Is Rattling Governments and Redefining Modern Journalism*. Cambridge, MA: Westview Press.

FCC ought to rope, brand CATV, Western TV operators tell Hill. 1958. *Broadcasting*, June 2, p. 60.

Federal Communications Commission, 2000. Fact sheet: Cable television information bulletin. June. Available from: www.fcc.gov/mb/facts/csgen.html. [Cited January 12, 2007]

Freeman, M. 2000. *ESPN: The Uncensored History*. Dallas: Taylor Publishing.

Goodwin, A., 1992. *Dancing in the Distraction Factory: Music Television and Popular Culture*. Minneapolis: University of Minnesota Press.

Gould, J., 1950. TV aerial on hill aids valley town. *New York Times*, December 22, p. 30.

Hard decision FCC faces now on pay cable, 1973. *Broadcasting*, November 12, p. 23.

HBO makes deals for movies, cable system, 1976. *Broadcasting*, June 28, pp. 55–6.

Hearn, T. and Farrell, M., 2006. FCC approves Adelphia merger. *Multichannel News*, July 13. Available from: http://www.multichannel.com/article/CA6352881.html. [Cited January 12, 2007]

Hilmes, Michele, ed., 2003. *The Television History Book*. London: BFI.

Howard, H. and Carroll, S., 1980. *Subscription Television: History, Current Status, and Economic Projections*. Knoxville: University of Tennessee Press.

Hoye, Jacob, ed., 2001. *MTV Uncensored*. New York: Penguin.

Johnson, N., 1967. CATV: Promise and peril. *Saturday Review*, 11 (November), pp. 87–8.

Johnson, R., 2003. Oral history interview transcript. Interviewed by B. Lamb. Denver, CO: The Cable Center. Available from: http://www.cablecenter.org/education/library/oralHistoryDetails.cfm?id=124. [Cited January 12, 2007]

Kagan, P., 1999. Oral history interview transcript. Interviewed by J. Keller. Denvern, CO: The Cable Center. Available from: http://www.cablecenter.org/education/library/oralHistoryDetails.cfm?id=128. [Cited January 12, 2007]

Kilgannon, C., 2006. After luck with poker, ESPN bets on New York dominoes. *New York Times*, April 2, pp. 1+ [Sec. 1].

LaMonica, P., 2006a. Google to buy YouTube for $1.65 billion. CNNMoney.com, October 9. Available from: http://money.cnn.com/2006/10/09/technology/googleyoutube_deal/. [Cited January 12, 2007]

LaMonica, P., 2006b. Meet the sons of YouTube. CNNMoney.com, July 24. Available from: http://money.cnn.com/2006/07/21/technology/onlinevideo/index.htm. [Cited January 12, 2007]

Laybourne, G., 2000. Oral history interview transcript. Interviewed by S. Nelson. Denver, CO: The Cable Center. Available from: http://www.cablecenter.org/education/library/oralHistoryDetails.cfm?id=134. [Cited January 12, 2007]

Le Duc, D., 1987. *Beyond Broadcasting: Patterns in Policy and Law.* New York: Longman.

Linder, L., 1999. *Public Access Television: America's Electronic Soapbox.* Westport, CT: Praeger.

Lockman, B. and Sarvey, D., 2005. *Pioneers of Cable Television.* Jefferson, NC: McFarland.

Long, M., 1985. *Satellite TV.* Mendocino, CA: Quantum Publishing.

Lucas, E., 1951. How TV came to Panther Valley. *Radio & Television News*, March, pp. 31–4+.

Magnant, R., 1977. *Domestic Satellite: An FCC Giant Step.* Boulder, CO: Westview.

Mair, G., 1988. *Inside HBO: The Billion Dollar War Between Hollywood and the Home Video Revolution.* New York: Dodd, Mead & Co.

Malarkey, Martin F., 1985. Oral history interview transcript. Interviewed by Max Paglin, 1985. (Denves, Co: The Cable Center. Available from: www.cablecenter.org. [August]

McPhail, T., 2006. *Global Communication: theories, Stakeholders, and Trends,* 2nd ed. Malden, MA: Blackwell.

Merina, A., 1993. Meet: John Hendricks. *NEA Today*, September 9.

Miles, H., 2005. *Al-Jazeera: The Inside Story of the Arab News Channel That Is Challenging the West.* New York: Grove Press.

Mullen, M., 1999. The pre-history of pay cable television: An overview and analysis. *Historical Journal of Film, Radio, and Television*, 19:1 (March), pp. 39–56.

——, 2003. *The Rise of Cable Programming in the United States: Revolution or Evolution?* Austin: University of Texas Press.

Myhren. T., 2000. Oral history interview transcript. Interviewed by P. Maxwell. Denver, CO: The Cable Center. Available from: http://www.cablecenter.org/education/library/oralHistoryDetails.cfm?id=150. [Cited January 12, 2007].

New Decade Productions in association with Channel Four Television, 1990. *Public Access, Everyone's Channel.* [Video] Bray Ireland: New Decade Productions. [David Shulman, Producer–Director–Writer]

Paglin, M. et al., eds. 1999. *The Communications Act: A Legislative History of the Major Amendments 1934–1996,* Silver Springs, MD: Pike & Fischer.

Parsons L., 1986. Oral history interview transcript. Interviewed by R. Barton. Denver, CO: The Cable Center. Available from: http://www.cablecenter.org/education/library/oralHistoryDetails.cfm?id=247. [Cited January 12, 2007]

Parsons, P., 1996. Two tales of a city: John Walson, Sr., Mahanoy City, and the "founding" of cable TV, *Journal of Broadcasting & Electronic Media* 40:3 (June 1), pp. 354–65.

Parsons, P. and Frieden, R., 1998. *The Cable and Satellite Television Industries.* Boston: Allyn & Bacon.

Phillips, M., 1972. *CATV: A History of Community Antenna Television.* Evanston, IL: Northwestern University Press.

Pine, B. and Gilmore, J., 1999. *The Experience Economy: Work Is Theatre & Every Business a Stage.* Boston, MA: Harvard Business School Press.

Rabinovitz, L., 1989. Animation, postmodernism, and MTV, *The Velvet Light Trap*, 24 (Fall), pp. 99–112.

Rifkin, M., 1998. Oral history interview transcript. Interviewed by J. Keller. Denver, CO: The Cable Center. Available from: http://www.cablecenter.org/education/library/oralHistoryDetails.cfm?id=252. [Cited January 12, 2007]

Robichaux, M., 2002. *Cable Cowboy: John Malone and the Rise of the Modern Cable Business.* Hoboken, NJ: John Wiley.

Rohm, W., 2002. *The Murdoch Mission: The Digital Transformation of a Media Empire.* New York: JohnWiley.

Rosenbloom, J., 1999. Cable Television Amendments. In Paglin, M. et al., eds. 1999. *The Communications Act: A Legislative History of the Major Amendments 1934–1996*, Silver Springs, MD: Pike & Fischer, pp. 213–382.

Saada, M., 1951. Stretching television: New "utilities" deliver TV to towns outside usual reception range. *Wall Street Journal,* January 3, p. 1.

Seiden, M., 1965. *An Economic Analysis of Community Antenna Television Systems and the Television Broadcasting Industry.* Washington, DC: GPO.

——, 1972. *Cable Television U. S. A.: An Analysis of Government Policy.* New York: Praeger.

Shapp, M. J., 1986. Oral history interview transcript. Interviewed by D. Phillips. Denver, CO: The Cable Center. Available from: http://www.cablecenter.org/education/library/oralHistoryDetails.cfm?id=261. [Cited January 12, 2007]

Sinclair, J., 1999. *Latin American Television: A Global View.* New York: Oxford University Press.

Sinclair, J. and Turner, G., eds., 2004. *Contemporary World Television.* London: BFI.

Sky Angel TV answers call for family-friendly, multi-channel television programming, 2005. *Business Wire,* June 23. Available from: http://www.findarticles.com/p/articles/mi_m0EIN/is_2005_June_23/ai_n13828268. [Cited January 12, 2007]

Sloan Commission on Cable Communications, 1971. *On the Cable: The Television of Abundance.* New York: McGraw-Hill.

Smith, A., ed., 1998. *Television: An International History.* New York: Oxford University Press.

Smith, E. S., 1987. Oral history interview transcript. Interviewed by P. Parsons. Denver, CO: The Cable Center. [Unavailable online]

Smith, R., 1972. *The Wired Nation*. New York: Harper-Colophon.

Southwick, T., 1998. *Distant Signals: How Cable TV Changed the World of Telecommunications*. Overland Park, KS: Primedia Intertec.

Sterling, C. and Kittross, J., 2002. *Stay Tuned: A History of American Broadcasting*, 3rd ed. Mahwah, NJ: LEA.

Streeter, T., 1997. Blue skies and strange bedfellows: The discourse of cable television. In L. Spigel and M. Curtin, eds., *The Revolution Wasn't Televised: Sixties Television and Social Conflict*, New York: Routledge, pp. 221–42.

Tarlton, R., 1993. Oral history interview transcript. Interviewed by J. Keller. Denver, CO: The Cable Center. Available from: http://www.cablecenter.org/education/library/oralHistoryDetails.cfm?id=173. [Cited January 12, 2007]

Taylor, A., 2000. *History Between Their Ears: Recollections of Pioneer CATV Engineers*. Denver: Cable Center.

Television & Cable Factbook, Annual. Washington, DC: Warren Communications News.

Thomas, A., 2005. *Imagi-Nations and Borderless Television: Media, Culture and Politics Across Asia*. Thousand Oaks, CA: Sage.

Thussu, D., 2006. *International Communication: Continuity and Change*, 2nd ed. London: Hodder Arnold.

US Federal Communications Commission (FCC), 1959. *Report and Order*. Inquiry into the Impact of Community Antenna Systems, Translators, TV Satellite Stations and TV "Repeaters" on the Orderly Development of Television Broadcasting. Washington, DC: GPO.

——, 1965. *First Report and Order*. [On CATV] Washington, DC: GPO.

——, 1966. *Second Report and Order*. [On CATV] Washington, DC: GPO.

——, 1969. *Fourth Report and Order*. [On pay-television] Washington, DC: GPO.

——, 1972. *Cable Television Report and Order*. Washington, DC: GPO.

Walker, J. and Ferguson, D., 1998. *The Broadcast Television Industry*. Boston: Allyn & Bacon.

Walson, J., 1987. Oral history interview transcript. Interviewed by M. Mayer. Denver, CO: The Cable Center. Available from: http://www.cablecenter.org/education/library/oralHistoryDetails.cfm?id=270. [Cited January 12, 2007]

Waterman, D., 1986. The failure of cultural programming on cable TV: An economic interpretation, *Journal of Communication*, 36:3 (Summer), pp. 92–107.

Wharton, D., 1995. Devotion to docs pays off, *Variety*, April 3–9. [Special supplement: Discovery: Ten Years of Exploring Your World], pp. 3–4+.

Whiteside, T. Onward and upward with the arts: Cable I-III, *New Yorker*, May 20, pp. 45–85; May 27, pp. 43–73; June 8, pp. 82–105.

Whittemore, H., 1990. *CNN: The Inside Story*. Boston: Little, Brown.

Index